FLY NAVY

D1345150

FLY NAVY

NAVAL AVIATORS AND CARRIER AVIATION – A HISTORY

PHILIP KAPLAN

AURUM PRESS

For my old friends Michael O'Leary and Malcolm Bates,
and in memory of Wing Commander Geoffrey Page

First published in Great Britain 2001 by
Aurum Press Ltd
25 Bedford Avenue
London WC1B 3AT

Text copyright © 2001 by Philip Kaplan

Page 252 is an extension of this copyright page

A catalogue record for this book is available from the British Library.

ISBN 1 85410 767 4

1 3 5 7 9 10 8 6 4 2
2001 2003 2005 2004 2002
Printed and bound in Singapore by Imago

CONTENTS

GLOSSARY

AAA: Anti-aircraft artillery; also called ack-ack or flak.

Ace: Fighter pilot credited with at least five aerial kills.

ACLS: Automatic Carrier Landing System.

ACM: Air Combat Manouvering.

AEW: Airborne Early Warning.

Air Boss: Officer responsible for carrier flight operations.

Air Wing: The complement of aircraft on a carrier.

Angle of Attack: The angle of the wing of an aircraft to the flight path of the aircraft.

ASW: Anti-Submarine Warfare.

AV-8B: Boeing McDonnell Douglas version of BAe Sea Harrier V/STOL jet fighter.

Avenger: Grumman TBF torpedo bomber.

Ball: Primary optical landing aid on a carrier; also called FLOLS, MOVLAS, or LLD.

Bandit: Identified enemy aircraft.

Barrier Net or Rig: A nylon webbing arrangement that can be set up to halt an aircraft whose arresting hook has malfunctioned.

Bat turn: Extreme sharp turn.

Bear a hand: Help out.

Bearcat: Grumman F8F fighter.

Belay that!: Cancel the previous message.

Bingo Field: A land runway where an aircraft can be diverted if the pilot is unable to land on a carrier.

Bird Farm: Nickname for an aircraft carrier.

Blue Angels: US Navy flight demonstration team.

Blue Shirts: Flight deck personnel who handle the chocks and chains, drive tractors, and operate the elevators.

Blue Water Ops: Flight operations flown beyond the reach of a Bingo airfield.

Boat, the: How pilots refer to the aircraft carrier.

Bolter: A failed landing on a carrier, when the aircraft arresting hook does not engage any of the four arresting cables on the deck.

Box: The channel on a catapult entry in which an aircraft nosewheel is positioned to facilitate the hook-up and launch.

Brown Shirt: The airplane captain responsible for the care and readiness of the aircraft.

Buccaneer: British Aerospace/Blackburn strike aircraft.

Burner: The afterburner unit on an aircraft jet engine, used to provide maximum thrust or power.

CAG: Carrier Air Group.

Call sign: Pilot's nickname.

CAP: Combat Air Patrol.

Catapult: A steam-powered system utilized to launch aircraft to flying speed from a carrier deck.

CATCC: The Carrier Air Traffic Control Center.

Cat grip: Grab handle on canopy jamb of F/A-18 and F-14 aircraft; also called the "towel rack".

CCA: Carrier Controlled Approach where a pilot is talked down the approach to the flight deck by a radar controller in the ship.

CDC: The Combat Direction Center on a carrier, where all tactical information is displayed and analysed.

Chock: A frame that is placed around an aircraft tire to prevent the plane rolling.

Chopper: Helicopter.

Chow: Food; a meal.

CO: Commanding Officer.

COD: Carrier Onboard Delivery, a twin-engined aircraft used to ferry cargo or passengers to and from an aircraft carrier.

Corsair: Chance-Vought F4U fighter.

Corsair II: Vought A-7 fighter.

Coupled Approach: A landing approach in which the autopilot of the aircraft is linked to radio navigation aids.

Cranial: The safety helmet worn by aircraft carrier flight deck personnel.

Cross-deck pendant: The segment of the arresting cable which is stretched across the flight deck and is engaged by the tailhook of an aircraft.

Crusader: Chance-Vought F-8 fighter.

Cut, the: LSO signal for pilot to land.

CV: US Navy designation for a conventionally-powered aircraft carrier.

CVE: US Navy designation for an escort aircraft carrier.

CVL: US Navy designation for a light aircraft carrier.

CVN: US Navy designation for a nuclear-powered aircraft carrier.

Cyclic Ops: Aircraft launches and recoveries from a carrier at one or two hour intervals.

Dauntless: Douglas SBD dive bomber.

Dilbert Dunker: A device for teaching emergency escape from an aircraft down in the sea.

Dropline: A series of orange lights down the stern centerline of a carrier which a pilot on approach uses to gauge the corrections he makes in his approach.

ECM: Electronic Counter-measures; a system used for the jamming of enemy detection, weapons and communications.

Eject: Emergency escape from an aircraft by means of an explosive-charged seat.

Elevator: An ascending and descending section of the flight deck of a carrier, used to transport aircraft between the hangar deck and the flight deck.

Etendard: Dassault Breguet strike fighter; also Super Et.

EWO: Electronics Warfare Officer.

FDO: Fighter Director Officer.

Fish: Torpedo.

Flag Plot: Admiral's working spaces.

Flak: See AAA.

Flanker: Sukhoi Su-27 jet fighter.

Flattop: Aircraft carrier.

Flight deck: The platform exposed surface on a carrier where aircraft are launched and recovered.

Float coat: Inflatable vest worn by flight deck personnel.

FLOLS: The Fresnel Lens Optical Landing System that is employed as the primary landing aid on an aircraft carrier; also called the Ball.

Fly-by-wire: The control surfaces of an aircraft are operated by electronic rather than mechanical or hydraulic means.

FOD: Foreign Object Damage which normally results when an aircraft engine ingests debris from the flight deck of a carrier.

FOD Walk-down: Carried out before every flight event; in which many personnel form a line across the flight deck, to walk its length in search of any objects that might foul aircraft engines or injure people if blown around by taxiing aircraft.

Fouled deck: When the flight deck of a carrier is not in a condition to receive recovering aircraft for any reason

FRS: Fleet Replacement Squadron; see RAG.

Fulcrum: MiG-29 jet fighter.

Galley: Where chow is prepared.

Gannet: Fairey ASW/AEW aircraft.

Gedunk: Term referring to a ship's store where candy and junk food is available. Also refers to ice cream.

Glidepath: The approach course of an aircraft that is descending and recovering to the flight deck of a carrier.

Glideslope: An optical or electronic beam that marks the approach glidepath of a recovering aircraft.

GQ: General Quarters.

Green Shirts: The flight deck crewmen who secure aircraft to catapults for the launch to flying speed.

Greyhound: See COD.

G-suit: A flight suit that brings compressed air pressure to reduce effects of gravity load forces on an aircrewman's lower body.

Handler: The person who is responsible for positioning, moving and arranging aircraft on the flight deck and hangar deck of a carrier.

Hangar deck: The deck

area of a carrier, under the flight deck, where aircraft are maintained and stored.

Hawkeye: Grumman E-2 airborne early warning plane.

Helix: Kamov Ka-31 RLD helicopter.

Hellcat: Grumman F6F fighter.

Helldiver: Curtiss SB2C dive bomber.

Helo Dunker: Device for teaching emergency escape technique from a helicopter down in water.

Holdback: A breakable link device which restrains an aircraft on the catapult until the instant of the launch.

Hook: The arresting tailhook of an aircraft, located beneath the tail, and utilized to engage an arresting cable on the flight deck of a carrier.

Hornet: Northrop Grumman F/A-18 jet fighter multi-role aircraft.

Hot-pumping: Refuelling an aircraft with its engine(s) running.

HUD: Head Up Display where the instrument data of an aircraft are projected onto a transparent screen between the pilot and the windshield.

ILS: Instrument Landing System.

INS: Inertial Navigation System.

Intruder: Grumman A-6 attack aircraft.

Island: A superstructure on the right side of the flight deck of a carrier, housing the bridge, PriFly and other personnel.

Jet blast deflector: A metal rectangular barrier that is raised behind an aircraft on the catapult awaiting launch,

to protect deck personnel.

Jinking: Evasive aircraft manoeuvre.

Joy Stick: Aircraft control column; often includes firing switches and other controls.

Kamikaze: World War II Japanese suicide pilot, attack, or aircraft.

Knee-knocker: The oval-shaped doorway through a bulkhead on a carrier.

Launch: When a catapult flings an aircraft from the flight deck of a carrier, accelerating it to flying speed.

Liberty: Shore leave.

LLD: Landing Light Device; the Ball on a carrier Fresnel Lens landing system.

LSO: The Landing Signal Officer, who is responsible for guiding the recovery of aircraft to the deck of a carrier. He or she judges and comments on aircraft approaches, assisting the pilots, giving grades and critiques of all landings; also called "paddles".

MAD: Magnetic Anomaly Detector device utilized to locate submerged enemy submarines.

Mae West: Inflatable life vest.

Main deck: On a carrier, it is the hangar deck; the deck from which the numbering of all other decks begins.

Marshal: Or marshaling stack; a holding pattern located behind the carrier, for aircraft prior to their approach and recovery.

Meatball: See Ball.

Military power: Maximum aircraft engine power.

Mouse: Self-contained two-way radio headphones for the use of certain flight deck personnel.

MOVLAS: The Manually

Operated Visual Landing System, used as a back-up for the FLOLS on a carrier.

Mule: A small tractor used for moving aircraft on the flight and hangar decks.

NAS: Naval Air Station.

NFO: Naval Flight Officer.

Nose tow: A tow bar that extends from the nosewheel of an carrier aircraft, which secures to the catapult for launching.

Now Hear This!: Listen up.

Nugget: Neophyte pilot.

O Club: Officer's club.

Old Salt: Experienced sailor.

Panther: Grumman F9F fighter.

Phantom: McDonnell Douglas F-4 fighter-bomber.

Pickle: A hand-held switch device by which an LSO can illuminate a row of red lights on the Ball to order a wave-off of an approaching pilot.

Port: Left side of ship. Also a harbour or docking place.

Precision Approach Radar: Shows the altitude and position of an aircraft that is approaching the carrier for a landing recovery.

PriFly: Primary flying control station in the island.

Primary Air Controller: See Air Boss.

Prowler: Grumman EA-6 attack/airborne early warning aircraft.

Purple Shirts: Deck crews responsible for refuelling of aircraft; also called "grapes".

Qual: Pilot's aircraft carrier qualification flights.

Rack: Enlisted man's bed; a three-high bunkbed.

RAG: Replacement Air Group; see FRS.

Ramp: The aft edge of the flight deck on a carrier.

Ready Room: Briefing and

lounge space for a squadron.

Red Shirts: Deck crews who handle ordnance, load bombs and ammunition; they also handle fire equipment.

Rig for: Set up or prepare.

RIO: Radar Intercept Officer who rides the back seat of an F-14 Tomcat jet fighter.

Roof: Slang for flight deck.

Rounddown: The ramp or aft end of the flight deck.

SA: Situational awarness.

SAR: Search and Rescue.

Scuttlebutt: Rumours.

Seafire: Supermarine fighter.

Sea Fury: Hawker fighter.

Sea Harrier: BAe V/STOL jet fighter.

Sea Hawk: Hawker fighter-bomber.

Seahawk: Sikorsky UH-60 helicopter.

Sea Hornet: DeHavilland fighter.

Sea Hurricane: Hawker fighter.

Sea King: Westland/Sikorsky Helicopter.

Sea Vampire: DeHavilland fighter.

Sea Venom: DeHavilland fighter-bomber.

Sea Vixen: DeHavilland fighter.

Shuttle: Device on catapult which protrudes above the flight deck; the aircraft nose tow is attached to the shuttle prior to launch.

Sick Bay: Ship's hospital.

Silver Suits: Handle aircraft crashes and fires.

Skyhawk: Douglas A-4 attack aircraft.

Skyraider: Douglas AD attack aircraft.

Slider: A hamburger.

SNJ: North American trainer.

Sortie: One mission flight by one aircraft.

Spool up: Running up the

power on a jet engine.

Squawk Box: Ship's intercom system.

Starboard: Right side of ship.

Strafe: To fire an aircraft's guns at a surface target.

Swordfish: Fairey multi-role bi-plane.

TACAN: Measures bearing and distance between the carrier and an aircraft.

Tail-end Charlie: The last aircraft in a formation.

Tomcat: Northrop Grumman F-14 jet fighter multi-role aircraft.

Torpecker: Torpedo bomber.

Tracer: Visible fired rounds.

Trap: An arrested landing of an aircraft on a carrier deck.

Turbo–Mentor: Beechcraft T-34 Primary trainer aircraft.

Val: Aichi D3A dive bomber.

VERTREP: Vertical replenishment at sea using helicopters.

VFR: Visual Flight Rules; operating procedures for use in good weather.

Viking: Lockheed S-3B ASW aircraft. Also ES-3A Shadow.

VSTOL: Vertical Short Take-Off and Landing.

Vulture's Row: Observation area located high on the island structure.

Wave-off: Aborted landing approach; an order from the LSO to a pilot to abort his landing approach.

White Shirts: Deck crewmen responsible for inspecting aircraft at the catapult prior to launch.

Wildcat: Grumman F4F fighter.

XO: Executive Officer.

Yellow Shirt: Crewman who directs the movement of aircraft on the flight deck.

Zero-Sen: Mitsubishi A6M fighter; A6M2 called Zeke.

ON 2 DECEMBER 1908 Rear Admiral W.S. Cowles, Chief of the US Navy Bureau of Equipment, recommended to the Secretary of the Navy that "a number of aeroplanes be purchased to operate from a ship's deck, carry a wireless telegraph, operate in weather other than a dead calm, maintain a high rate of speed and", among other things, "be of such design as to permit convenient stowage on board ship." Thus began the history of the aircraft carrier and naval aviation.

At the advent of the aircraft carrier, the major navies of the world were battleship forces. These great vessels were the largest, most heavily armed and armoured warships afloat, and their very presence had on occasion seen them prevail without even engaging the enemy, so intimidating were they. Why then did the aircraft carrier virtually scuttle the venerable battlewagon as the new capital ship of the world's great sea powers? To begin with, it was her versatility. The aircraft carrier is at once a self-propelled mobile airbase that is essentially self-sufficient, as well as being a wholly effective servicing facility. Going into the twenty-first century, the modern US supercarrier operates as the spearhead of an awesome force called a carrier battle group. The other types of ships in carrier battle groups — destroyers, frigates, cruisers, replenishment tankers and fast attack submarines — are there to protect "mother", the CVN or nuclear-powered aircraft carrier. The carrier weapon has another edge over the battleship. It gives a fleet the ability to fight and win a naval battle at arm's length, out of sight of the enemy. At the start of the new millennium, the twelve operational carrier battle groups of the US fleet, and its assault carriers, are capable of covering immense areas of ocean, sending massive air strikes at inshore as well as ocean targets, providing air support for amphibious landings, helicopter troop ferrying, search-and-rescue, and conducting highly sophisticated anti-submarine warfare. While perhaps less than totally

THE SHIP

Wake, friend, from forth thy lethargy; the drum / Beats brave and loud in Europe, and bids come / All that dare rouse, or are not loath to quit Their vicious ease and be o'erwhelmed with it. It is a call to keep the spirits alive / That gasp for action, and would yet revive / Man's buried honour in his sleepy life, / Quickening dead nature to her noblest strife. / All other acts of worldlings are but toil / In dreams, begun in hope, and end in spoil.
— from *An Epistle to a Friend, to perswade him to the Warres* by Ben Jonson

Left: The USS *Saratoga* (CV-3) in the 1930s. The *Lexington*-class ships (*Lexington* and *Saratoga*) are considered the first truly effective aircraft carriers of the US Navy. They had Special Treatment Steel hulls two inches thick and were the most powerful US warships of World War II.

invulnerable to enemy attack, the CVN, within the protective shelter of her battle group, is the biggest, widest-ranging, most powerful and fearsome threat that has ever put to sea. In just a quarter of a century her capabilities have utterly eclipsed those of her battleship predecessor. According to the late United States Senator John C. Stennis (namesake of the USS *John C. Stennis*, CVN-74), Chairman of the US Senate Armed Services Committee from 1969 to 1980, and a strong supporter of a powerful, well-trained and well-equipped military, "The best way to avoid war is to be fully prepared, have the tools of war in abundance, and have them ready." Of the modern nuclear-powered aircraft carrier, Senator Stennis said: "It carries everything and goes full strength and is ready to fight or go into action within minutes after it arrives at its destination . . . there is nothing that compares with it when it comes to deterrence."

Right: HMS *Illustrious* is the fifth Royal Navy ship to bear that name. The 20,000 ton *Invincible*-class aircraft carrier can support a mix of up to 22 aircraft and is a sister ship to *Invincible* and *Ark Royal*.

Clément Ader, a less than successful early French aviator, made the world's first "flight" in a powered airplane from level ground on 9 October 1890. He did the hop in a steam-powered monoplane, the *Eole*, a large bat-like craft. Ader was then asked by the French Minister of War to design, build and test a two-seat plane to carry a light bomb load for the military. The result was a failure. It crashed at Satorg on 14 October 1891 and the contract for the war plane was cancelled. Ader, however, had an unshakeable faith in his personal vision of the future of aviation. He accurately forecasted that land warfare would be transformed by the practice of reconnaissance from aircraft, and that it would also revolutionize the methods of war fleets at sea. He predicted that they would carry their aircraft to sea with them. He coined the term *porte-avions* (aircraft carrier) and wrote that such ships would be unlike any other, with clear, unimpeded flight decks, elevators to take aircraft (with their wings folded) from the flight deck to stowage below for servicing and repairs and bring them back again, and the ability to operate at a high rate of speed. His foresight was amazing.

The world's first reasonably successful take-off by a fixed-wing aircraft from the deck of a ship was made by Eugene Ely, an exhibition pilot for aircraft designer and builder Glenn Curtiss, achieved in a Curtiss Pusher in November 1910. Ely flew the craft down a gently sloping wooden platform on the forecastle of the US light cruiser *Birmingham* on the afternoon of 14 November as the ship steamed slowly in Chesapeake Bay. In the attempt, Ely's aircraft actually hit the water once, but he retained control and managed to land safely on the shore. In another equally significant trial held at San Francisco in January 1911, Ely successfully landed a Curtiss aircraft on a specially-built platform over the quarterdeck of the cruiser *Pennsylvania*. He used undercarriage hooks to engage one of twenty-two transverse wires that were stretched across the platform and anchored at the sides with sandbags. Within months of this second successful carrier trial, Eugene Ely was dead, killed in an air crash.

In the month prior to Ely's landing trial, Curtiss offered at his own expense, "to instruct an officer of the US Navy in the operation and construction of a Curtiss aeroplane." On 23 December Lieutenant T.G. Ellyson reported to North Island in San Diego bay, and four months later was "graduated" by Curtiss who wrote to the Secretary of the Navy: "Lt. Ellyson is now competent to care for and operate Curtiss aeroplanes."

Within eight years of the Wright brothers' initial powered flight at Kitty Hawk on the North Carolina coast, the capability to take off from and recover an aircraft to a carrier deck (using arresting wires) was proven. Gradually the American Navy became convinced of the value and importance of aviation for patrol and reconnaissance in its

Right: Pioneering aviator
Eugene Ely and one of
the Curtiss Pusher
aircraft that he operated
in his early aircraft
carrier demonstrations.

future. It concluded that floating airports —
aircraft carriers — would have to be developed
and acquired to support and exploit its combat
airplanes. However, it was the British who, by the
occasion of the Royal Navy's 1913 Naval
Manoeuvres, had identified and defined nearly all
the fundamental requirements for carrier-borne
operation, aircraft and equipment. Wing-folding to
facilitate the improved stowage of otherwise bulky
aircraft on the limited space of a ship deck, the
testing and evaluation of bomb dropping, gun
mounting and firing, and the successful launch of
a 14-inch torpedo from early British-built carrier-
borne aircraft, propelled the Royal Navy to the
forefront of such development. They also
concluded that the purpose-built aircraft carrier
must be developed and, by the end of World War I,
had achieved that goal. The way to this
achievement led to the 1914 conversions of the
bulk carrier *Ark Royal* and the large light cruiser
Furious to aircraft carriers, *Ark Royal* being
designed almost from scratch to meet the needs
of naval air operations as they were then perceived.
At the same time the Japanese began work
converting the merchant ship *Wakamiya* to carry
seaplanes. The French were actually first to make
such a conversion when, in 1912, they changed a
torpedo depot ship to an operational seaplane
carrier.

The American navy followed the British lead and
converted the collier *Jupiter* into an experimental
aircraft carrier, recommissioning it as the USS
Langley on 20 March 1922. On 27 December of
that year Japan commissioned its first aircraft
carrier, the *Hosho,* and the first take-off from its
deck was performed in a Mitsubishi 1MF1 by a
British pilot named Jordan. *Hosho* was one of the
most significant aircraft carriers ever built.
Extremely small, displacing a mere 7470 tons, she
was so well-designed, her spaces so well-planned,
that she could operate up to twenty-six aircraft —
a remarkable capability for her size. Even more

remarkable, however, was the pioneering
installation of an experimental light-and-mirror
system for aiding pilots in their landing approaches.
It was a concept far ahead of its time, as the
approach speeds of the carrier aircraft of the day
and for many years to come were far slower than
those which were to require the eventual design
and refinement of the ultimate mirror light landing
system. It is ironic that *Hosho* should be the only
one of the ten principal carriers with which Japan
fought World War II, to survive that war.

During World War I the Royal Navy, motivated by
German airship bombing raids on East Coast
Britain and reconnaissance activity over the North
Sea, leaned towards development of sea-borne
aircraft in an early warning/interception role:
floatplanes operating from carrier vessels in the
North Sea. But the concept was flawed. The
seaplanes were simply too slow and lacked an
adequate rate of climb, preventing them from a
timely closure with the German aircraft. The drag
created by the floats on the underpowered British
planes was just too great. Still, the German airship
campaign compelled the Britons to stick with the
ineffectual concept.

A Royal Navy pilot, Squadron-Commander E.H.
Dunning, was operating aboard HMS *Furious* in
the summer of 1917. The ship was extremely fast
and Dunning was convinced that her speed would
provide the key to developing a routinely safe
landing procedure for his pilots, who were flying
the agile Sopwith Pup. Dunning soon discovered
that the Pup could be easily and gently coaxed
over the flight deck of *Furious* to an acceptable
landing. The ship had no arrester wires, but
Dunning had rope handles fitted to his Sopwith
aircraft and instructed the deck hands to grip the
handles once he had landed, to hold the little
plane firmly on the deck. On 2 August 1917 he
approached *Furious* and made the first successful
carrier deck landing on a ship that was steaming
into wind. In a similar landing attempt two days

Early Royal Navy carriers.
Below left: HMS *Glorious*.
Below right: HMS *Argus*.
Bottom: HMS
Courageous.

later, Dunning touched down on the deck, a tyre on the Sopwith burst, and the aircraft cartwheeled off the edge and into the sea before the deck party could stop it. Dunning was drowned. It was soon accepted that, for an aircraft carrier to operate its aircraft effectively and with relative safety for the pilots and aircrew, a much larger hull would have to be utilized to provide a far greater deck area — an area closer to that of a conventional airfield.

The warship designers had to rethink their approach to planning the layout of an aircraft carrier deck and the necessary bridge, funnel and other external structures. Consensus finally led to acceptance of a full-length flight deck, as free of obstructions to the safe launch and recovery of the ships' aircraft as could be achieved. Safety and practicality required that take-offs be made from the forward end of the ship, and landings onto the aft end. Naval architects had to wrestle with the

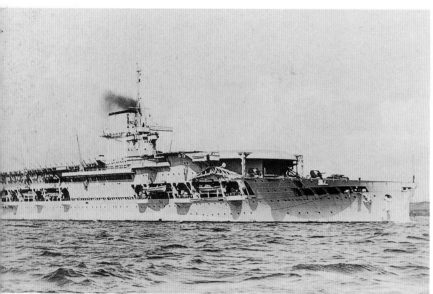

considerable challenge of positioning the ship's funnel in such a way as to at least minimize the hazard to aviators caused by the smoke of the ship's boiler gases which reduced visibility for the airmen on landing. Various ideas were tried over many years of carrier development, including a system of ducting to take the smoke discharge towards the stern rather than allowing it to eminate from amidships, as was the norm in island superstructure placement. The problem of smoke from a carrier funnel flowing across the flight deck, as it sometimes did, was never really resolved until the advent of the (smokeless) nuclear powerplant for aircraft carriers in the 1960s. Logic and pilot preference eventually dictated the starboard placement of the island superstructure, adopted by most of the world's navies for their carriers. The verticality of the narrow island structure which became a standard, and the funnel positioned behind and in line with

Below: The launching on 20 March 1945 of USS *Midway* (CVA-41). A major refit for *Midway* in the 1960s enabled her to operate modern jets from her greatly enlarged and angled flight deck.

Below: With hinged stacks, the first American aircraft carrier, the USS *Langley* (CV-1) at North Island, San Diego *circa* 1932. The *Langley* was modified from the US Navy fleet collier *Jupiter*. Her normal complement of aircraft was twelve scouts and ten torpedo bombers.

it, stems from the need to put the bridge, and especially the navigation platform, relatively high and to discharge the smoke conveniently, if not always to the advantage of landing aviators.

In 1927 the new US Navy carriers *Lexington* and *Saratoga* became the platforms on which the transverse arrestor wire-landing technique was perfected. Their 34-knot high speed and other features fulfilled Clément Ader's prophecy.

The period from 1928 through the decade of the

1930s was, despite the economic depression and the drastic restrictions on most nations' economies, one of important development of carriers and their aircraft and tactics. Britain got to grips with the threat posed by Germany re-arming, and began a radical re-evaluation of the needs of her armed forces. The Fleet Air Arm had been operated under the control of the Air Ministry and had not been well managed under these auspices. It lacked state-of-the-art aircraft and had lost many of its Royal

Naval Air Service pilots through absorption into the newly-formed Royal Air Force in 1918. Inter-service rivalry between the Royal Navy and the RAF finally caused the British government in 1937 to return the responsibility for British naval aviation to the Royal Navy.

Among the most important contributing factors to the development of the aircraft carrier by the great naval powers of the world during the 1930s, was the Washington Naval Treaty of 1922. The agenda at Washington was clearly to redefine the sea power of the signatories, in part by imposing a strict rationing on heavy warship construction in a blatant attempt to minimize battleship building. That ultimately worked, though not through the honourable adherence of the treaty participants to the type and tonnage limitations they had imposed on themselves. Rather, future development of battleships was largely curtailed owing to a clause in the treaty which allowed the signatories to opt for converting their unfinished battleship hulls to completion as aircraft carriers. Britain opted to complete her unfinished battleships *Nelson* and *Rodney* as battleships, which later proved to be a mistake as they were unable to achieve more than a 23-knot top speed in a time when aircraft carriers were already capable of speeds in excess of 30 knots. The United States and Japan elected to take up the carrier option, and in so doing were able to launch fast new carriers in the same year that *Nelson* and *Rodney* were launched. In another controversial decision, the British chose to convert *Furious* to a proper aircraft carrier, but in the effort they chose to give her a flush-deck in combination with a short bow flying-off deck, for reasons of economy as much as utility. They gave the same design to her sister ships *Glorious* and *Courageous*. The Japanese, who for several years had been heavily influenced by British carrier design and innovation, promptly based the design of their newest carriers *Akagi* and *Kaga* on that of

Furious. The mistake here lay mainly in not recognizing that carrier-borne aircraft, both of the future and those about to come on stream, were becoming larger, heavier and faster, and were certainly going to require considerably more flight deck for take-offs and landings, and more stowage space. The bow flying-off deck would quickly become obsolete. After many years of leading the field in aircraft carrier design and innovation, the British were suddenly no longer on the same page as the Americans.

Like many pilots who fly for airlines today, Frank Furbish came from a military aviation background. He was on the USS *Constellation* (CV-64) from 1984 through to 1986, flying the F-14A with VF-21. "Life on an aircraft carrier is like no other. More than 5000 men (and women now) are packed tightly within the floating city. As large as the ship is, the spaces are generally small and are separated by watertight doors. Only two levels allow continuous fore and aft access — the second deck and the 03 level. Even those levels have 'knee-knockers' every few yards. On the *Constellation* there are two stores with sundry items, a barber shop and a small gym. Room accommodations are directly related to one's rank. Enlisted personnel are in large bunk rooms. Junior officers are in smaller bunk rooms, and senior officers either have their own rooms or share a room with another officer. I had eight roommates. Our junior officer bunk room was about the size of a large hotel room, with three sets of three-deep bunks which were nothing more than metal frames and foam pads. The spacing between the bunks was so restrictive that big-shouldered guys could not roll over without first jutting out of their bunks. We had a desk/locker combination, a few stand-up gym lockers and one sink. Privacy was obtained by hanging a curtain around your bunk. Your room was not only your rest area, but also your office. There was little air-conditioning and plenty of

We sail the ocean blue
And our saucy ship's a beauty
We're sober men and true
And attentive to our duty.
– from *HMS Pinafore*
by William S. Gilbert and
Arthur Sullivan

Out of a fired ship, which, by no way / But drowning, could be rescued from the flame, Some men leap'd forth, and ever as they came / Near the foe's ships, did by their shot decay; / So all were lost, which in the ship were found, / They in the sea being burnt, they in the burnt ship drowned.
– *A Burnt Ship*
by John Donne

noise. Rest was constantly interrupted by the needs of fellow roommates to report for watches at all hours of the night. Rest and paperwork was also affected by the location of the berthing. Aircrews are required to be on the 03 level, just below the flight deck. The noise level there is quite high during flight ops, with 30-ton aircraft landing just above your head. Additional noise was provided by the 'chock chuckers' throwing heavy metal chocks and tie-down chains, and by 'paint chippers'. It always amazed me that there was a constant need for someone to chip paint somewhere on the ship.

"The bathroom or 'head' was another unique feature. On a conventional (non-nuclear) ship like the *Constellation*, the water always had the after taste of fuel oil. The showerhead was a hose with a knob. The knob had a button that had to be held in for the 'mist' to dispense, in order to save water, which fluctuated between scalding and ice cold. At times it would just quit. This usually happened after you had covered yourself with soapy lather. It seems trivial, but after six months of such showers the desire for a long, warm shower with uninterrupted water is enormous.

"On another deployment I graduated to a four-man bunk room located forward on the port side just under a catapult and abeam of the water-break. The water-break was the device that stopped the catapult shuttle, which travelled at 150 m.p.h. as it tossed a multi-ton aircraft into the sky. Imagine, if you will, that your cot is inside a dumpster and someone is randomly crashing a large sledge hammer into the side of it. You simply cannot sleep, period. Some people believed that if you could manage to sleep twelve hours a day the cruise would only be half as long. Some tried."

During the Korean War Bill Hannan served aboard the USS *Kearsarge* as a jet engine mechanic. "It would be difficult to imagine a more hazardous

The freedom of the seas is the *sine qua non* of peace, equality and co-operation.
– from an address to the United States Senate, 22 January 1917 by U.S. President Woodrow Wilson

An aircraft carrier is a noble thing. It lacks almost everything that seems to denote nobility, yet deep nobility is there.
  A carrier has no poise. It has no grace. It is top-heavy and lop-sided. It has the lines of a well-fed cow.
  It doesn't cut through the water like a cruiser, knifing romantically along. It doesn't dance and cavort like a destroyer. It just plows. You feel it should be carrying a hod, rather than wearing a red sash.
  Yet a carrier is a ferocious thing, and out of its heritage of action has grown nobility. I believe that today every navy in the world has as its No. 1 priority the destruction of enemy carriers. That's a precarious honor, but it's a proud one.
– Ernie Pyle

Above left: A US Navy aviation fueller at work on a carrier. Above centre: A USN ordnance man between tasks. Bottom left: An F-14 Tomcat two-seat multi-purpose fighter and an F/A-18 Hornet on a US carrier in the 1990s. Left: A plane director spotting an F/A-18 on the number two catapult of an American carrier.

Left: The USS *Midway* (CV-41) shares a pier with two US cruisers. At upper left is the battleship USS *Missouri* (BB-63), on whose deck the Japanese signed the surrender document ending World War II. Above: An F6F Hellcat about to land on the *Independence*-class carrier, USS *Langley* (CVL-27) in 1944.

place to live than an aircraft carrier. Apart from the airplanes landing (and sometimes crashing) on our roof, the ship carried millions of gallons of flammables, including fuel oil, high-octane gasoline, paints, thinner, as well as tons of bombs, rockets and assorted ammunition. And scuttlebutt had it that we were being shadowed by submarines.

"Although much of our maintenance work was performed on the hangar deck, some of it was topside, under incredibly harrowing conditions. Imagine working on a wintry night, on a frosty deck in rough seas, close to the edge of the ship. Only dim, red-lens flashlights could be employed under blackout conditions. If a man happened to slip over the side into the ocean, about 80 feet below, there was little chance of him being missed, let alone rescued. Being assigned to plane-pushing meant long hours of physically hard work in particularly dangerous circumstances. Once during my turn on this dreaded duty, I was well forward on the wet flight deck during a launch operation when I was suddenly blown to the deck. I went sliding rapidly towards the propeller planes poised aft with their engines running. I was desperately clawing at the wooden deck, hoping to grab onto one of the numerous metal tie-down strips, when a huge bruiser of a fellow spotted my predicament and, at considerable risk to himself, literally tackled me and dragged me over the edge of the deck into a catwalk. Probably no one else had even seen me. I was speechless, and didn't even have a chance to thank him before he rushed back to his post . . . a hero in my view, but probably all in a day's work on the flight deck for him.

"Oral communication was often difficult on the flight deck, even without aircraft engines running. With jet and prop plane run-ups, it was almost impossible. The ship's powerful 'bull horn' public address system could sometimes be heard over the din, but, even in the absence of noise, the bull horn messages were sometimes garbled or distorted like old-time railroad station announcements. So an

Believe me, my young friend, there is *nothing* – absolutely nothing – half so much worth doing as simply messing about in boats.
– from *The Wind in the Willows*
by Kenneth Grahame

The gloomy hulls, in armour grim, Like clouds o'er moors have met, / And prove that oak, and iron, and man / Are tough in fibre yet. / But Splendours wane. The sea-fight yields / No front of old display; / The garniture, emblazonment, / And heraldry all decay.
–from *The Temeraire*
by Herman Melville

Left: HMS *Illustrious* driving through a squall in the Bay of Biscay in 1999. Originally intended as purely submarine-hunting vessels, she and sister ships *Invincible* and *Ark Royal* have evolved into strike-fighter platforms since the Falklands War in 1982. Overleaf: The 81,773-ton *Kitty Hawk*-class carrier USS *Constellation* (CV-64) departing San Diego bay.

Clockwise from top left: Stanley Vejtasa, former skipper of the USS *Constellation* (CV-64). A captain's cap from the USS *John C. Stennis* (CVN-74). A catapult crewman during the Gulf War. Night launch of an F/A-18 Hornet from a USN nuclear-powered supercarrier. The Royal Navy's HMS *Invincible*, to the left of the French carrier *Clemenceau*, which was withdrawn from service at the end of the 1990s. Together with her sister ship *Foch*, *Clemenceau* operated in the Mediterranean from the 1970s. In 1990 she brought helicopters and troops to the Gulf for the Allied Coalition conflict with Iraq after the Iraqi invasion of Kuwait.

elaborate system of hand-signals evolved that usually worked to make a point.

"The *Kearsarge* converted sea water into fresh quite efficiently, but it had to serve many purposes, the first of which was feeding the ship's thirsty boilers. Conservation measures had to be strictly observed. The shower protocol, for example, was to wet down, turn off the water, soap up, turn on the water, rinse, and turn off the water. Once, I had just finished soaping up and was covered with suds, when 'general quarters' was sounded (loud horns over the intercom system, meaning 'get to your duty station immediately!') I grabbed a pair of shower sandals and my jockey shorts and rushed up to the hangar deck. A 'red alert' was in effect and we could not leave our stations for any reason until the all-clear was sounded. Still covered in soapsuds, I was very cold and beginning to itch all over. But it could have been worse. At least I wasn't up on the flight deck.

"Field Days were clean-up times, and meant hard work to anyone involved. The worst thing about them was the closure of the heads (toilet facilities) for cleaning. Time on the ship for 'pit stops' was at a premium there, and often there was little margin for delay in attending to such bodily functions. How frustrating then, after climbing down a deck or two, to find a HEAD SECURED FOR CLEANING sign on the entrance, and, either by coincidence or design, any nearby heads were also frequently out of service.

"We headed home at a leisurely pace, probably to reduce wear and tear on the ship's machinery. As keeping idle hands out of mischief was a foremost consideration in the Navy, all sorts of chores, such as chipping paint from decks and then repainting them, were assigned to anyone who appeared to be unoccupied. In our squadron, the most unsavory of our F9F Panthers was selected to have its paint manually stripped, a dirty, tedious job at best, especially when it was freely admitted that the job didn't really need doing."

Rear Admiral Dennis Campbell, CB, DSC, RN died on 6 April 2000 at the age of 92. Campbell's career as a Fleet Air Arm aviator exposed him to many near-misses and close calls. He realized early on that one of the greatest hazards faced by a naval aviator was the probability of a collision with parked aircraft (or a barrier placed in front of such aircraft to protect them) if he should fail to catch an arresting cable in his deck landing. He came up with an amazingly simple solution . . . the angled flight deck, which allowed carrier pilots to approach the deck at a slight angle to the ship itself. Should the pilot fail to catch one of the arresting cables, he could "bolt" and go around again for another landing approach by applying full power on touch-down and flying off the clear port bow. The idea was so successful that both the Royal and US Navies adopted it immediately and most large aircraft carriers now have an angled deck for recovering aircraft. Thanks to Dennis Campbell (and to the gyro-stabilised mirror landing system which continuously signals glide path corrections), today's naval aviator is able to make a landing approach on a constant speed and direction, and with a clear escape route should his tailhook fail to engage an arresting cable.

SHE WAS THE "Stringbag" to all who flew in her or were associated with her. The Fairey Swordfish was a remarkably successful aircraft in its principal role as a carrier-based torpedo bomber, and performed creditably in other roles as well. She resulted from specs for a naval torpedo bomber, issued in July 1932 to the Fairey Aviation Technical Department. The requirement called for an aircraft to perform:

1. Fleet reconnaissance with a crew of three
2. Torpedo attacks with a crew of two and of limited range, and
3. Bombing attacks with two or three crew according to the bomb load.

Trials of several early models of the plane were ultimately to result in specification S.38/34 calling for three developmental aircraft to be built. The subsequent production machines based on these craft became the final Swordfish.

John W.G. Wellham flew the Swordfish in the Fleet Air Arm attack on Taranto in November 1940: "It was astonishingly easy to fly. It was never used as a trainer because it was too easy to learn on. It had a very low take-off speed, a very low landing speed and was incredibly robust. You could thump it down on a carrier deck, you could knock bits off it, you could bend it, and it still went on flying. It would carry almost anything you put on it. It was originally supposed to carry torpedos and a few bombs. Gradually they added more and more: a modern radio, radar, later the homing torpedo, depth charges and, if we were travelling somewhere, as when the squadron made a sudden move, we even tied motor bikes to the torpedo rack, spare propellers to the wings, and stuffed everyone's luggage into the back. It still took off and flew. Technically it wasn't allowed, but you could do aerobatics in it. I once did a loop with a roll off the top of it, and it did it perfectly happily."

"A venomous-looking little bumble bee of a fighter, the Wildcat — destined to be the progenitor of a

HARDWARE

Problem: Something loose in cockpit. *Solution*: Something tightened in cockpit.

Problem: Evidence of leak on right main landing gear. *Solution*: Evidence removed.

The Crusader was designed in 1953 as a carrier-based fighter that was also capable of supersonic speeds. To achieve that previously unattainable feat, the Vought F-8 aircraft incorporated a variable-incidence wing which could be pivoted upwards at an angle of up to 7°. This feature gave the Crusader lower approach speeds to the deck, and greater lift for take-off, while affording excellent visibility for the pilot. Many Crusaders were in action in Vietnam, with a final tally of eighteen MiG-21 aircraft downed.

Left: Aboard the USS *Hancock* (CV-19) an A-1B Skyraider of VA-215 is given the "turn-up engine" signal prior to launch by the catapult officer of the carrier which was operating from Yankee Station off the coast of Vietnam in May 1966.

The old Able Dog, Straight Six [Skyraider] (not the Spad!) was the Navy's only good atomic bomber during the fifties. It could carry an extremely heavy load off a carrier, when no other type, excepting maybe the high-flying AJ Savage, could do the job. The AD had incomparable range. It could loiter and return. We might have been slower than the F9F-8 Panthers, but we flew farther. When Navy carrier jets were made into attack types in the 1950s, they could not carry atomic bombs, and it was said that in a war, they had such short range that they would have to fly one-way missions. In our air group, the other attack squadron flew the F9F-8. I was never more proud of my AD than the day CAG (Carrier Air Wing Commander) required the F9F Panther squadron to demonstrate how it would carry a 'shape' off the cat. A shape was an inert object of the same size and weight as an atomic bomb. I watched as the Panther attempted to get catapulted while carrying a shape. The pilot sat ready on the cat, with full thrust blasting away, converting fuel into noise, while the ship made its best knots and searched for wind. The pilot waited to launch, but within minutes he had used so much fuel that he would not have reached the simulated target, and the exercise was cancelled.
– Paul Ludwig, former US Navy attack pilot

dynasty of 'Cats' — made up for its lack of aesthetic appeal in purposefulness of appearance. . ." wrote Royal Navy test pilot Eric Brown of the Grumman plane that fought against the best the enemy had in 1942. The Japanese Zero could easily out-perform the F4F Wildcat in a turn, climb or dash, and was only at a disadvantage against the Grumman fighter when the Wildcat had both height and surprise on its side. In a fight, the Wildcat had one other thing going for it. Though out-gunned by the Zero, the Grumman was better able to absorb damage than the Japanese plane, and the Wildcat pilot was far better protected by armour and a bulletproof windscreen.

In the Atlantic the Royal Navy operated the Wildcat, which they called the Martlet (though later in the war they came to accept the US nickname), from small escort carriers which sailed with Allied convoys and provided essential air cover protecting against enemy aircraft and U-boat attacks. Among the best-known achievements employing the Wildcat is that of Lieutenant Edward H. O'Hare of VF-42, flying from the USS *Lexington* (CV-2) on 20 February 1942. The *Lexington* had launched the Wildcat squadron to intercept inbound Japanese aircraft and Lt. O'Hare managed to down five Betty bombers that day, for which he was awarded the Congressional Medal of Honor.

Bob Croman flew the SBD Douglas Dauntless, referred to by some of its crews as "the clunk" or "the barge". In its role as a dive-bomber, Croman recalls: "We dropped bombs, diving from up to 15,000 feet. We pulled out at about 1500 feet and in this performance, it [the SBD-4] never failed me. The only minor fault I found was that the landing flap handle and the dive flap handle were next to each other and occasionally one could move the dive flaps when landing. This resulted in many accidents. The SBD was very capable in carrier landings and I very seldom received a wave-off from the LSO." Eric Brown states:

"Mundane by contemporary performance standards, the Dauntless was underpowered, painfully slow, short on range, woefully vulnerable to fighters, and uncomfortable and fatiguing to fly for any length of time, being inherently noisy and draughty. But it did possess certain invaluable assets that mitigated these shortcomings. Its handling characteristics were, for the most part, innocuous and it was responsive; it was dependable and extremely sturdy, capable of absorbing considerable battle damage and remaining airborne and, most important, it was an *accurate* dive bomber.

"Throughout the Pacific War, it remained the principal shipboard dive-bomber available to the US Navy; it was the only US aircraft to participate in all five naval engagements fought exclusively between carriers, and, deficiencies notwithstanding, it emerged with an almost legendary reputation as the most successful shipboard dive-bomber of all time — albeit success that perhaps owed more to the crews that flew it in truly *dauntless* fashion than to the intrinsic qualities of the aeroplane itself."

The Curtiss SB2C Helldiver was the third in a line of Curtiss planes to carry that name. The SB2C was purpose-built to be both a bomber and dive-bomber, and was a sturdy, two-man, low-wing, all metal monoplane which had a production run of 7200. Flown by both the US Navy and Marines, the later versions of the Helldiver were armed with either four 12.7mm wing-mounted machine guns, or two 20mm cannons. Late variants were also produced by both Fairchild, and Canadian Car & Foundry, which was a manufacturer of later marks of the Hawker Hurricane fighter as well.

The Imperial Japanese Navy sent its requirements for a new front line fighter to both the Nakajima and Mitsubishi aircraft companies in 1937. It wanted a top speed of at least 310 m.p.h., the ability to climb to 10,000 feet in 3.5 minutes, exceptional manoeuvrability and greater range than that of any existing fighter. It had to be armed with two cannons and two machine guns, a requirement thought to be unrealistic by both manufacturers, and Nakajima pulled out of the competition to design and build it. Although Mitsubishi agreed to solve the problem of the plane's armament, the requirement was eventually modified. In their determination to keep the weight of the new fighter to a minimum, thus enhancing its overall performance, a new lightweight alloy which the manufacturers called Extra-Super Duralumin (ESD) was used. The new carrier fighter was called Zero-Sen, and designated A6M1. It was accepted by the Imperial Navy on 14 September 1939. Later, when the Navy accepted a new 925 hp engine built by Nakajima, the NK1C Sakae 12, the next Zero variant carried the new powerplant and was designated A6M2. It was this greatly improved aircraft that Allied fighter pilots

Below: F/A-18 Hornets of the US Navy Flight Demonstration Team, The Blue Angels, over Florida. In the new century, the Hornet is still the Navy's primary state-of-the-art strike fighter, and will be until the new Joint Strike Fighter is operational.

in the Pacific had to contend with in 1943 prior to the arrival in that theatre of their own new and highly capable Corsair and Hellcat fighters.

Among Japan's principal performers in their surprise attack on the US battleships in Pearl Harbor, which brought America into World War II, was the Aichi D3A Val naval dive-bomber. A successor to Aichi's D1A, the Val was prominent in Japan's strikes throughout South-East Asia and the South Pacific until late in 1942 when the quality and performance of the US carrier fighters improved dramatically. After that, the Val was no longer in the game and was utilized in secondary roles, including training, land-based attack and, finally, in the desperate suicide attacks of 1945 against the American fleet.

No one has flown and evaluated more military aircraft types than Eric Brown, the extraordinary test pilot who became the first naval officer to head the élite Aerodynamics Flight at the Royal Aircraft Establishment, Farnborough, England: "As a result of a brilliant piece of improvisation, the Navy had been presented with the Sea Hurricane which had proved that a high-performance shore-based fighter *could* be operated with relative safety from a carrier, but the Hawker fighter's chances of survival against a Bf 109G or Fw 190 were anything but good. Nevertheless, its successful adaptation for the shipboard role had at least brought about something of a revolution in naval thinking, and logically enough in 1941 the Admiralty began to demand a similar adaptation of what was then the highest performing fighter available — the Spitfire.

"This scheme was received with mixed feelings by those naval pilots in the know. Everyone admired the Spitfire and itched to fly it — but from an aircraft carrier? That was a horse of a very different colour! No one needed convincing of the performance or handling attributes of this magnificent fighter which was surely one of the greatest warplanes ever conceived, but there was a certain air of fragility about the aeroplane: a ballerina-like delicacy that seemed inconsistent with the demanding, muscle-taxing scenario of shipboard operations. Could that slender fuselage stand the harsh deceleration of an arrested landing? Would that frail undercarriage absorb the shock of 15 ft/sec (4. 57 m/sec) vertical velocity and could those wafer-like wings take the acceleration forces of a catapult launch? What of the Spitfire's high landing speed, but, above all, would the pilot ever be able to see the carrier deck on the approach?

"I had been selected to undertake the first Seafire deck trials on an escort carrier. At 1330 hours on 11 September 1942 I was flying a Seafire towards HMS *Biter* — the time being significant as it turned out. As I approached *Biter*, I confidently assumed that all would be breached up for this momentous occasion — the first landing of a Seafire on one of these postage stamp-sized platforms — and I did a quick circuit at 400 feet (120 m) and settled into my approach. Contrary to earlier recommendations, I adopted a straight final approach with the aircraft crabbed to starboard so that I had a good, clear view to port of both the deck and, supposedly, the batsman of whom, in the event, I saw no sign.

"Nevertheless, all went smoothly and I touched down on *Biter*, picking up the second wire. I came to a stop somewhat smartly and cut the engine, and it was only then that I realized that the deck was completely deserted! Then striding from the island came a lone figure — the Captain. He stepped up on the wing and said, with a pleasant smile, 'I say old boy, you were not expected so early and everyone's at lunch!' It was only then that the truth dawned on me: I had landed with the carrier 25 degrees out of wind, only 13 knots (24 km/h) windspeed over the deck, the arrester wires lying flat and unsupported, and no batsman!"

In World War II Grumman Aircraft Engineering

Corporation of Bethpage, New York quickly established itself as the premier designer and manufacturer of high-quality, dependable and effective carrier-borne aircraft. The tradition continued with the coming of the TBF/TBM Avenger torpedo-bomber to the American fleet in 1942, and to the British Fleet Air Arm early in 1943. Well-armed and quite capable of defending itself, the structurally rugged Avenger proved to be a superb warplane. Operating primarily from escort carriers, FAA Avengers flew anti-submarine patrol missions and mine laying sorties over the waters surrounding Britain, and worked well in the campaign to deny the use of the English Channel to enemy shipping in the run-up to the Normandy landings. FAA Avengers were extremely effective against the Japanese in the Pacific war, flying from the carriers *Illustrious, Victorious, Indefatigable, Indomitable* and *Formidable*, and attacking enemy airfields and oil refineries. The big Grumman performed magnificently and inspired confidence in its crews. It is regarded as one of the

Below: A Royal Navy Martlet II (Wildcat) of No. 888 Squadron aboard HMS *Formidable* in the summer of 1942. Two Martlet pilots of 888 combined to down a Kawanishi flying boat in the Bay of Bengal on 2 August for the type's only victory over Japanese aircraft in that year.

The Hellcat had a kill ratio of 19 to 1; the P-51 had a kill ratio of 14 to 1, and the Corsair had a kill ratio of 11 to 1. Captain [Eric] Brown, Royal Navy test pilot, flew all the enemy and allied fighters of the war. He said that the Hellcat and the Fw 190 were the two best fighters. I never flew the Fw 190, but flew the Hellcat 396 hours in combat. To say I love the Hellcat is an understatement. Only my family do I love more.
– Richard H. May, former US Navy fighter pilot

Right: The powerful and formidable Grumman Hellcat fighter proved its worth hundreds of times during the Pacific Theatre air fighting in World War II. There is no greater showcase of the plane's capability than the 19 June 1944 air battle remembered as "The Marianas Turkeyshoot". The US Navy Hellcats performed magnificently in the action and, thereafter, the air arm of the Imperial Japanese Navy was no longer a significant threat to Allied forces in the Pacific.

best shipboard aircraft of World War II.

Of what may have been the most effective US Navy fighter of that war, Eric Brown wrote: "No more outstanding example of skill and luck joining forces to produce just the right aeroplane is to be found than that provided by the Grumman Hellcat, unquestionably the most important Allied shipboard fighter of World War II. US Navy observers . . . had learned that speed, climb rate, adequate firepower and armour protection, pilot visibility and manoeuvrability were primary requirements in *that* order. In the case of the shipboard fighter, these desirable qualities had to be augmented by ample fuel and ammunition capacity, and the structural sturdiness that was a prerequisite in any aeroplane intended for the naval environment. The Grumman team had . . . formulated its own ideas on the primary requirements in a Wildcat successor, these having been arrived at by consensus: the opinions of US Navy pilots had been canvassed and the results equated with analyses of European combat reports.

". . . so spacious was the cockpit that when I first lowered myself into the seat, I thought I would have to stand up if I was to see through the windscreen for take-off! There was a great deal of engine ahead, but in spite of the bulk of the Double Wasp, the view was not unreasonable, thanks to the Hellcat's slightly hump-backed profile. The take-off was straight forward, with a tendency to swing to port requiring gentle application of rudder, and acceleration was very satisfying, although the noise with the canopy open was quite something. In a dive it was necessary to decrease r.p.m. to 2100 or less in order to reduce vibration to acceptable levels. The Hellcat had to be trimmed into the dive and became tail heavy as speed built up, some left rudder being necessary to counter yaw. With bombs under the wings, speed built up very rapidly indeed and there was considerable snatching of the elevator and rudder. Final approach to the deck

was at 80 knots and the Hellcat was as steady as a rock, with precise attitude and speed control. I always used a curved approach to get a better view of the deck and left the throttle cut as late as possible to prevent the heavy nose dropping and allowing the main wheels to contact the deck first, thus risking a bounce, although the Hellcat's immensely sturdy undercarriage possessed very good shock absorbing characteristics.

"The Hellcat had twice the power of the [Japanese] Zero-Sen, but it also carried twice the weight. The Japanese fighter was supremely agile and its turning radius was fabled. At airspeeds below 200 knots, the Zero-Sen's lightly loaded wing enabled it to outmanoeuvre the Hellcat with comparative ease, even though the airframe of the latter allowed more g to be pulled. At higher speeds the controls of the Japanese fighter stiffened up and the turn rate disparity was dramatically reduced.

"The Hellcat could not follow the Zero-Sen in a tight loop and US Navy Hellcat pilots soon learned to avoid any attempt to dog-fight with the Japanese fighter, and turn the superior speed, dive and altitude characteristics of the American fighter to advantage. These characteristics usually enabled the Hellcat to acquire an advantageous position permitting the use of tactics whereby the Zero-Sen could be destroyed without tight-manoeuvring combat.

"The Hellcat had not been the fastest shipboard fighter of World War II, nor the most manoeuvrable, but it was certainly the most efficacious, and one of its virtues was its versatility. Conceived primarily as an air superiority weapon with the range to seek out the enemy so that he could be brought to battle, as emphasis switched to offensive capability, the Hellcat competently offered the strike potential that the US Navy needed to offset the reduced number of bombers aboard the service's fleet carriers. The Hellcat was certainly the most important Allied shipboard aircraft in the Pacific in 1943–44, for it turned the tide of the conflict, and

Problem: Number three engine missing.
Solution: Engine found on right wing after brief search.

Problem: Aircraft handles funny.
Solution: Aircraft warned to straighten up and fly right.

Right: In Grumman's Bethpage, Long Island, New York Plant One, several F4F-4 Wildcats in early 1942 markings in the final stages of assembly. The larger, twin-engined aircraft in the foreground is a JRF-6B Goose amphibian in British markings.

although Japanese fighters of superior performance made their appearance, they were too late to wrest the aerial ascendancy that had been established largely by this one aircraft type."

In the World War II era, the US Navy had a pronounced preference for powering its fighter aircraft with radial air-cooled engines, while the US Army preferred liquid-cooled inline engines for its fighters. The Navy believed in the ruggedness and reliability of the radial engine and thought it better suited to long over-water flights. It was natural then that Chance–Vought selected the huge Pratt & Whitney R-2800 to power their unusual and impressive new fighter, the F4U Corsair. Called "the bent-winged bird" and "whistling death", the Corsair came from the drawing board with many inherent flaws which were to require more than a hundred separate modifications relating to lateral stability and aileron control alone. Still, when the XF4U-1 prototype flew in 1940, the Navy realized it was on course towards the special air superiority fighter it needed so desperately. The prototype surpassed 400 m.p.h. in level flight, the first American combat airplane to do that. Teething troubles with the engine challenged Pratt & Whitney for a time, but the marriage of the powerplant to the F4U airframe was a good one, ultimately producing one of the finest propeller-driven airplanes of all time.

A radical "gull-wing" design for the Corsair was the Chance–Vought solution to the problem of propeller clearance necessitated by the 13' 4" prop needed to transfer power from the R-2800 engine. With a conventional wing design, the airplane would have an unacceptably high angle of attack when landing or taking off. The new wing design also allowed the pilot increased visibility across the wing at the "bend", an improved drag factor, which added to the performance, and the folding-point of the wing aided in the stowage problem on board carriers.

Eric Brown comments: "The incipient bounce

36

Right: Anti-Submarine Warfare is the primary responsibility of the Sea King helicopters of No. 814 Naval Air Squadron, Royal Navy Fleet Air Arm. The Royal Navy Sea King Mk 6 is one of the most versatile and effective large maritime helicopters in service today. Based on the American Sikorsky SH3 design, it is built entirely in the United Kingdom, using British systems and electronics. It is capable of fully independent day and night operations in all weather conditions and has achieved great success in search-and-rescue missions.

during landing resulting from overly stiff oleo action . . . was found to be most disconcerting by pilots fresh to the F4U-1, but not so disconcerting as the virtually unheralded torque stall that occured all too frequently in landing condition. Compounded by a serious directional instability immediately after touch-down — and coupled with the frightful visibility from the cockpit during the landing approach — it was small wonder that the US Navy concluded that the average carrier pilot was unlikely to possess the necessary skill to master these unpleasant idiosyncrasies, and that committing the Corsair to shipboard operations until these quirks had been at least alleviated would be the height of foolhardiness.

"Oddly enough, the Royal Navy was not quite so fastidious as the US Navy regarding deck-landing characteristics, and cleared the Corsair for shipboard operation some nine months before its American counterpart. The obstacles to the Corsair's shipboard use were admittedly not insurmountable, but I can only surmise that the apparently ready acceptance by their Lordships of the Admiralty of the Chance–Vought fighter for carrier operation must have been solely due to the exigencies of the times, for the landing behaviour of the Corsair really was bad, a fact to which I was able to attest after the briefest acquaintance with the aircraft.

"I was well aware that the US Navy had found the Corsair's deck-landing characteristics so disappointing in trials that it had been assigned for shore duties while an attempt was being made to iron out the problems, and although the Fleet Air Arm *was* deck-landing the aircraft, I knew that, by consensus, it had been pronounced a brute and assumed that shipboard operations with the Corsair were something of a case of needs must when the devil drives. The fact that experienced US Navy pilots *could* deck-land the Corsair had been demonstrated a couple of months earlier. I was most anxious to discover for myself if the

Corsair was the deck-landing dog that it was reputed to be. It was!

"In the deck-landing configuration with approach power, the Corsair could demonstrate a very nasty incipient torque stall with dangerously little warning, the starboard wing usually dropping sharply. With the large flaps fully extended the descent rate was rapid, and a simulated deck-landing at 80 knots (148 km/h) gave very poor view and sluggish aileron and elevator control. A curved approach was very necessary if the pilot was to have any chance of seeing the carrier, let alone the batsman! When the throttle was cut the nose dropped so that the aircraft bounced on its mainwheels, and once the tailwheel made contact the aircraft proved very unstable directionally, despite the tailwheel lock, swinging either to port or starboard, and this swing had to be checked immediately with the brakes. Oh yes, the Corsair could be landed on a deck without undue difficulty by an experienced pilot in ideal conditions, but with pilots of average capability, really pitching decks and marginal weather conditions, attrition simply had to be of serious proportions."

Eventually most of the negative characteristics of the Corsair were overcome and her pilots came to appreciate her excellent performance. In World War II Corsair pilots destroyed 2140 enemy aircraft for a loss of 935 Corsairs (190 in aerial combat, 350 due to anti-aircraft fire, 230 from other causes, and 165 in crash-landings). A further 692 were lost on non-operational flights.

Just too late for World War II, the Douglas A-1 Skyraider joined the US military inventories in 1946 and really began to show what it could do when the Korean war broke out in 1950. The Skyraider is easily the most unique, versatile and utilitarian single-engined propeller-driven combat plane of all time. Able to carry a bomb load greater in weight than its own empty weight, it was operated very successfully by several US Navy, US Marine and US Air Force squadrons, as well as the South Vietnamese Air Force during the Vietnam

In the 1960s a primary ship-based utility helicopter for the US Navy was the Kaman SH-2D and SH-2F Seasprite. It was to become, through a series of conversions, a highly sophisticated multi-purpose helicopter in the light airborne multi-purpose system role (LAMPS) in the early 1980s. Powered by two General Electric T58 turboshaft engines, the Kaman was a launch platform for AIM-7 Sparrow and AIM-9 Sidewinder air-to-air missiles and it was also armed with chin-mounted mini-gun turrets and waist-mounted machine guns for search-and-rescue operation. Some were equipped for anti-submarine and anti-ship missile defence work. With substantial range and relatively powerful radar, the Seasprite made a significant contribution as a reliable Navy workhorse.

Below: The full-load displacement of the Russian aircraft carrier *Kuznetsov* of 58,500 tons is roughly three times that of HMS *Illustrious*, and one-half that of the USS *John C. Stennis*. Her air wing includes a combination of twenty-two fixed-wing fighters and training planes, and up to eighteen helicopters. *Kuznetsov* incorporates both a ski jump take-off ramp and an angled flight deck for fixed-wing aircraft recovery.

conflict. It has also been in the service of Britain, France, Cambodia, Chad and the Central African Republic. In Vietnam it was referred to as "Spad" or "Sandy". The big 318 m.p.h. plane could carry 8000 pounds of ordnance on sixteen underwing pylons, considerably more than the bomb load of a four-engined B-17 bomber in World War II.

Among the many wonderful aircraft of the Duxford, England-based Fighter Collection is, arguably, the ultimate piston-engined air superiority fighter of the World War II era: the Grumman F8F Bearcat. Stephen Grey is the man behind The Fighter Collection and he flies the Bearcat, as well as several other splendid fighter types, in air shows every year, the greatest being Flying Legends each summer at Duxford. Grey probably knows the Bearcat at least as well as anyone who has ever had anything to do with it.

"One approaches the Bear with care. On the deck, it looks exactly as it was intended — the smallest airframe that could be mated to the Pratt & Whitney 2800 engine, whilst swinging that huge Aeroproducts propeller, 12' 7" in diameter, with the broadest blades you ever did see and just 6" ground clearance, without the oleos depressed. The whole thing is achieved by sitting it on extraordinarily long mainwheel legs. They have, of necessity, been attached to the extremities of the wing centre section and an ingenious double hinge system is adopted to fold the upper portion outwards as the main portion folds in. Difficult to describe but, believe me, an amazing amount of ironmongery folds away into a small hole (as the actress said to the Bishop). Somehow, they also found space for a 150 gallon centreline droptank.

"The span is only 35' 6", the length an exceptionally short 27' 8" but it sits 13' 8" high. A mean, muscular-looking, long-legged Bulldog. The cockpit is tiny, European style, with controls well placed but with canopy rails so tight against my shoulders that I have to be helped in — damn that apple pie.

"I am in it, so fly it. Those early days were quite educational. The only piston engined fighter I had then flown was the Mustang. Whilst the 51 has bags of torque, no one would complain that it was overpowered — rather, slightly overweight, like certain pilots I know. The briefing was pretty exhaustive: 'Do not exceed 54" and get the gear up before 140 knots.'

"Having been used to the long throttle travel on the 51 and needing all of it on take-off, I was surprised that an inch of throttle movement gave me 50" and a very rapid view of the side of the runway. I pulled up the nose to avoid exceeding the gear speed and the runway lights, to find myself in what seemed like the vertical. Finally cleaned up, I looked down to see this major Florida airfield as a tiny model, thousands of feet below. Out over the Everglades, even back at 1900 r.p.m. and 29.5", I was still several miles behind the aircraft. Within a few more minutes I had a sensation of familiarity and increasing exhilaration. The Bearcat is a joy to throw around. Harmony of control, aided by its spring tab ailerons, is outstanding; the controls delightfully light and very responsive. It demonstrates all the hallmarks of a great day fighter in that it has poor stability, which translates into rapid divergence and, with its high power/thrust to weight ratio, outstanding agility. The manual says this is not an instrument aeroplane. Believe it.

"Clean stalls are under the book figure of 100 knots, as the Fighter Collection aircraft is unarmed and lighter. The onset has little warning, excepting a very brief stick shudder prior to an immediate wing drop. With everything down, the stall comes at 70 knots, the nose dropping hard and violently off to the left after the same brief shudder. A rapid application of power will torque you around the prop. Potentially disastrous on the approach.

"Accelerated stalls are straightforward, until you try them whilst reducing or, still worse, adding coarse power. The resulting snap rolls have to be experienced to be believed. The aircraft is so short

coupled, you need the draught over the tail at any time you go for high Alphas.

"Another cute trick, which took me some time to work out, is aileron reversal. Induced by simultaneously pulling hard and rolling at very high speeds at low altitude, I first put in hard left aileron and found myself rolling right. Then, just as you have considered what the hell is happening and the speed is backing off, the Bear rolls hard left. We went right through the geometry, the rigging, the spring tabs, the lot, before I flew it again.

"Subsequent, deep examination of the manual showed a max deflection aileron limit of 4.5g. An experienced old salt confirmed that the wing was first twisting unloaded, then the ailerons work. This was, apparently, a 'design feature' of all Bearcats. Whoever said wing warping was ineffective? Needless to say, we now do no barrel-like manoeuvres at high speed in the Fighter Collection Bearcat.

"Another part of the early education was during practice for the first air show. I pulled up for the first Cuban and whilst inverted looked for the rollout target on the airfield, only to find it twice as small as the same in a 51. Yippee, I will do a one-and-a-half downward instead of a half. Through 360 degrees and back to inverted, the Bear was oscillating longitudinally and beginning to tuck under. A quick glance at the ASI showed I had hit the Candy bar, 435 knots and still accelerating. Fortunately I was unladen and still rolling (the correct way) when I hit the compressibility dive brakes. The oscillation stopped and I now had the problem of avoiding the increasingly large airfield. We thoroughly inspected the airframe and my underwear after that one.

"I do not wish to leave you with the impression of an impossible roaring monster. The aeroplane is a beautiful, well domesticated 'Pussycat' (but still with teeth and claws) when flown well within its envelope — as we have done since those early 'hooligan' days.

"The Bearcat was the end of an Era, the end of a

COD or Carrier Onboard Delivery aircraft have provided a precious link between aircraft carriers and land facilities since the days when modified World War II TBF and TBM Avengers were used in the role. Since then the dominant performers in that category have been the Grumman C-1 Trader and the Grumman C-2A Greyhound. The Trader was a twin-engined, prop-driven utility craft that was derived from the earlier Grumman S-2 Tracker anti-submarine aircraft. Deliveries to the US Navy began in 1955. With a 900-mile range, a speed of 287 m.p.h. and a cargo capacity of 8250 pounds, the C-1 (minus a cargo load) could also ferry up to nine passengers from beach to boat and back. The successor to the Trader was the Greyhound, a twin-engined turboprop aircraft whose design is based on that of the hugely successful Grumman E-2 Hawkeye airborne early warning and fighter direction aircraft. The Greyhound can seat thirty-nine passengers in a special configuration, and is capable of carrying small vehicles, spare jet engines and other aircraft parts. It fills the current COD requirement.

The excellent Sikorsky UH-60 Seahawk provides the US Navy with a near-perfect all-weather platform for its vital search-and-rescue and anti-submarine warfare missions. Capable of being airborne and on the job in just a few minutes, this amazing helicopter is as adept at finding and rescuing downed pilots as it is at locating enemy submarines which it can disable by dropping its own torpedoes.

technology. The ultimate piston engined air superiority fighter? The Spitfire XIV or the Sea Fury could give it a run but different fun."

Hawker's World War II Typhoon ground attack fighter led to their Tempest which, in turn, led to Sydney Camm's Project F.2/43, resulting in a great naval fighter, the Sea Fury, with deliveries of Sea Fury F Mk Xs to the Fleet Air Arm beginning in February 1947.

With the invasion of South Korea by the North Korean army in June 1950, the British Government contributed the services of the British Eastern Fleet headed by the carrier HMS *Triumph*. The carrier was to operate off Korea together with ships of the US Navy. HMS *Theseus* replaced *Triumph* in October, bringing with her No. 807 Squadron and its twenty-one Sea Fury FB Mk 11s. Subsequently, *Theseus* gave way to HMS *Glory*, as British carriers continued to operate off the Korean coast in six-month cycles. The Sea Fury proved highly successful as a fighter bomber in the demanding conditions of the Korean conflict. Perhaps its finest hour came on 9 August 1952 when four of the piston-engined fighters of 802 Squadron were intercepted by eight MiG-15 jets, resulting in a spectacular air fight and the downing of one MiG, with two others severely damaged. None of the Sea Furies were hit.

Since the 1960s, the Sea Fury has become popular with warbird enthusiasts, airshow and air racing participants and spectators, being among the fastest propeller-driven fighters of all time.

The US Navy was able to field the highly-effective Grumman F9F Panther fighter on more than 78,000 combat sorties in the Korean War because in Britain Rolls-Royce had developed an excellent jet engine in its Nene. The US Navy became interested in the Nene in the summer of 1946 when looking for an engine to power its new Grumman-designed, radar-equipped, all-weather day and night fighter, the XF9F. Two Nenes were brought from England to the Philadelphia Navy Yard where, in December, the Rolls product became the first jet to pass the extremely demanding Navy test schedule. Rights were acquired for American aircraft engine manufacturer Pratt & Whitney to redesign and build a version of the Nene to be designated J42, which was developed and refined to ultimately power the F9F2, the Panther that, together with the McDonnell F2H Banshee, did the primary day fighter job in Korean skies. The design evolved into the F9F-8 Cougar which, while not arriving in Korea in time to impact on the conflict, was to become the primary fighter of the US Navy in the fifteen years after World War II.

The Hawker Seahawk came about when the company elected to develop the fighter at its own expense, after receiving no British government interest in the project. As development progressed, however, the Ministry of Aircraft Production awoke to the plane's potential, ordering three prototypes in May 1946. The company proceeded with the project though, by then, what little interest the Royal Air Force may have had in the plane had evaporated, and the RAF opted instead for the new Gloster Meteor F. Mk 4. Emphasis at Hawker then shifted entirely to development of a navalized version to be Rolls-Royce powered. The Seahawk entered deck-landing trials on HMS *Illustrious* in 1949 and there followed full systems trials aboard HMS *Eagle* on the first nine aircraft from the initial production run of 151 machines. A ground-attack variant was also developed and by 1952 the Seahawk entered service trails. The front line version in 1956 was the Seahawk FGA Mk 6, then being operated by Royal Navy Fleet Air Arm squadrons aboard HMS *Albion*, HMS *Eagle* and HMS *Bulwark*. They effectively attacked Egyptian airfields and other military installations, and provided support for ground troops during the joint British-French

November Suez Operation, *Musketeer*.

The Royal Navy needed to replace the obsolescent DeHavilland Sea Hornet in 1950 and evaluated the DH Venom for that purpose. It wanted a navalized, carrier-borne two-seat nightfighter, and promptly ordered three prototype Sea Venoms. The requirement evolved into one for an all-weather fighter, and the first production model of the type flew initially in March 1953. Mk 20 Sea Venoms came into service in March 1954 with No. 890 Squadron, RN, at RNAS Yeovilton in Somerset, England. Undercarriage and arrestor hook weaknesses, however, required modification and it was July 1955 before the Sea Venoms could actually go to sea, aboard HMS *Albion* in the Mediterranean. Unfortunately, a series of spectacular landing accidents on the carrier (accidents related to further problems with the arrestor hook) caused the official clearance for Sea Venom carrier operations to be withdrawn. Significantly improved Sea Venom 21s and 22s became operational in the next few years, and the airplane proved extremely effective in Operation *Musketeer* during the Suez Crisis of 1956. Mk 21s of No. 809 Squadron in HMS *Albion*, and Nos. 892 and 893 Squadrons in HMS *Eagle*, while working in their primary role as interceptors, also supported the fleet and allied landings at Port Said. Using both cannons and rockets, the Sea Venoms hammered enemy airfield and military vehicle targets thoughout the six days of the operation. None of the aircraft were lost in the action, and only one was hit by anti-aircraft fire, forcing the pilot to make a wheels-up landing on *Eagle*. Sea Venoms were also operated by the French Navy and the Royal Australian Navy.

The McDonnell Douglas A-4 Skyhawk was rugged, reliable and quite agile, so much so that its carrier-based US pilots referred to it as "Scooter". It was almost always ready to work, and in the Vietnam war the nimble delta-winged jet could carry and deliver

The Hawker Seahawk entered service with the Royal Navy in 1953. Designed as a single-seat carrier-borne fighter, it had a Rolls-Royce Nene jet engine of 5200 pounds thrust. The Seahawk is best described as "dinky", with an empty weight of just 9560 pounds. It had very nice lines and flew superbly, indeed its handling was viceless. It handled so well that almost every squadron produced a four-aircraft formation aerobatic team.

All in all, the Seahawk was a flexible and effective fighter/ ground attack aircraft which saw service with the Royal Navy, the Royal Netherlands Navy, the Federal German Navy and the Indian Navy, where it remained in service until 1984 when they [the Royal Navy] took delivery of their Sea Harriers. The Seahawk was one of my favourite aircraft. In it I flew 780 hours and carried out 121 deck landings.
— Alan J. Leahy, former Royal Navy fighter pilot

Left: A 1950s ad from Hawker Aircraft Limited for their Sea Hawk jet carrier fighter.

nearly its own weight in bombs and rockets. The little Skyhawk was one of the many brilliant designs of Douglas Aircraft's Ed Heinemann, and was initially a private venture by Douglas to be a successor to their extraordinary Skyraider. At the time, the US Navy had in mind an aircraft of similar capabilities to those of the Skyraider, but powered by a turboprop engine. Heinemann had a better idea, and designed a relatively small airframe to be powered by a Pratt & Whitney turbojet and capable of the payload and range requirements of the Navy, with a far-greater-than-specified performance, and about half the specified maximum take-off weight. The result performed impressively for decades as a primary strike fighter for the US Navy, US Marines, Argentina and Israel.

The British Aerospace/Blackburn Buccaneer S. Mk 2B was rushed to the Persian Gulf in 1991 to provide laser designation for the missiles of RAF Tornados as they struck at Iraqi targets. The two-seat, low-level strike plane originally served with Nos. 800, 801 and 809 Squadrons, Royal Navy, in the S Mk 1 version. The brilliant, long-lived Buccaneer had been mobilized and was in actual combat in just ten days. The Bucs utilized an American system called the AVQ-23E Pave Spike daytime-only laser-designator pod, and a Westinghouse AN/ASQ-153 Pave Spike system, to help the Tornados accurately deliver laser-guided bombs from medium altitude. As the Gulf air war progressed, Buccaneers were also used to drop Portsmouth Aviation CPU-123/B 1000-pound Paveway II laser-guided bombs, mainly attacking bridges and airfields. The Buccaneer was among the most important contributions of the British to the coalition effort in the Gulf. Armed with AIM-9L Sidewinders for self-defence until the missiles were declared unnecessary, the Bucs alone made it possible for the RAF Tornado bomber force to hit its targets with unparalleled precision and efficiency. The Tornado crews had never dropped

laser-guided bombs before, and had to undergo a quick and very intense training program with their Buccaneer laser buddies prior to their first LGB mission on 2 February. From the start, the Buccaneer-Tornado operating procedure was a great success. For the thirty-year-old Buccaneer, fighting what would probably be its last action, it was an impressive performance.

Two training planes have had a more profound influence on naval aviators of the US and Britain than all other such aircraft combined in the history of navy flying — the North American T-6 SNJ Texan/Harvard, and the Beech T-34 Mentor. Probably more aviators of all branches have trained in the Texan/Harvard than in any other trainer. A demanding machine, it has been said of the T-6 that, if a pilot can learn to fly it properly, he can fly anything. Several other countries, including Canada which called the plane the Yale, have used it in their pilot training programs. The prototype first flew in April 1935. T-6s are still flown privately throughout the world. The plane is universally recognized as the most important trainer of World War II.

In 1948 the US Air Force was in the market for a new primary training plane and by 1950 it was evaluating the Beech YT-34, a basic trainer/light strike aircraft. The later T-34C variant, with a 550 hp Pratt & Whitney turboprop engine, has proven to be a superbly reliable, long-serving craft for both Air Force and Navy training needs, with deliveries beginning in 1977. All US naval aviators and Naval Flight Officers learn to fly the T-34 "Charlie" Turbo Mentor at NAS Pensacola or NAS Whiting Field in Florida, whether they go on to fly jets, fixed-wing propeller planes or helicopters. Referred to by Navy pilots as the "Turbo Weenie", the T-34C is relatively difficult to fly, which is just how the Navy wants it, in the belief that if a student can be taught to master the T-34, he or she will find flying a jet easy.

Known as "the airborne eyes of the fleet", the

The US Navy's H-46 Sea Knight was a similar, but much smaller version of the CH-47 Chinook helicopter, developed in 1959 to meet a Navy and Marine Corps requirement for a vertical replenishment (VERTREP) helicopter to airlift troops, parts, munitions and small vehicles from logistic support ships to combatant ships and beachheads. More than 1200 Sea Knights were produced including 134 under licence by Kawasaki in Japan. The US Marine Corps took delivery of 600 CH-46A aircraft starting in 1962. Roles for the Sea Knight included long-range search-and-rescue, and mine counter-measures.

Left: Five US Navy Grumman TBF Avenger torpedo bombers in a striking image. The big plane carried either one 22-inch torpedo or a 2000-pound bomb load.

The Grumman E-2 Hawkeye is a carrier-based airborne warning and control aircraft, powered by twin turboprop engines. It is operated by a crew of five. Weighing 55,569 pounds fully loaded, the Hawkeye has a top speed of 374 m.p.h. and a range of 1700 miles. It can remain airborne for seven hours unrefuelled. The service ceiling of the airplane is 30,000 feet. It has a radar detection range of more than 250 miles, and can track up to 300 targets and interpret 30 targets simultaneously. The aircraft originally cost US 32.4 million in 1978 dollars for a six-plane buy for the US Navy.

Grumman E-2 Hawkeye has for nearly four decades carried the primary responsibility for airborne early warning and operational co-ordination in the US Navy. It is expected to continue as the front line AEW and control platform for the American fleet well into the twenty-first century. Like so many naval aircraft designed and brought into manufacture in the 1950s and 1960s, the Hawkeye has been continually revised electronically to keep it technologically current. Subject to a prolonged early development phase owing to its complexity, the plane finally reached the Pacific and Atlantic fleets in quantity between 1964 and 1966. The Hawkeye flies with a five-man crew including a pilot, co-pilot and three radar operators, as a shipboard early warning and fighter control aircraft. Its two Allison turboprop engines power it to a maximum speed of 397 m.p.h. and a typical cruising speed of 315 m.p.h. with a ceiling of 31,700 feet and a ferry range of 1905 miles. Hawkeye first went to war aboard the USS *Kitty Hawk* (CVA-63) in the South China sea off Vietnam in 1965 where it began its highly successful career controlling air strikes, warning of enemy aircraft in the vicinity and guiding friendly aircraft around ground defence installations.

Plagued with problems from the outset, the Grumman (now Northrop Grumman) F-14 Tomcat remains a bittersweet combination of virtue and vice. Three decades old, the big interceptor still offers a unique threat to potential enemies of the US Fleet. It can destroy multiple targets at extreme distances. It can bring down enemy fighters before they are a threat, and missile-launching aircraft before they can launch their weapons. The Tomcat, the largest and heaviest of the American navy's strike aircraft, is showing its age and is gradually being replaced by the highly capable F/A-18 Hornet. It will continue to be replaced by the new Joint Strike Fighter when it comes on stream sometime before 2010. Not the most agile of air combat planes, the Tomcat was designed to defeat planes like the MiG-21 and the

MiG-23, but is not a match for the likes of the MiG-29 or the Su-27, and is certainly outclassed by the F-15 and F-16. The F-14 has suffered upwards of forty losses owing to spinning accidents, a slew of engine compressor section blade failures resulting in many more losses, and a growing lack of trust by its pilots in the Tomcat engines, bringing with it a collective attitude of excessive caution. The engine problems, however, were resolved with production of the F-14A, B and D models, and the Tomcat gained a new lease of life in the 1990s. The collapse of the Soviet Union and the end of the Cold War, and resultant heavy cutbacks in US military spending then brought a major contraction in the Tomcat community, as well as the disestablishment of many historic squadrons which were equipped with the plane. Some such squadrons are disappearing entirely, while others will ultimately convert to the F/A-18.

Lieutenant J.G. Jennifer Brattle, a Naval Flight Officer with VF-211, *Fighting Checkmates*, aboard the USS *John C. Stennis* (CVN-74), flies the F-14A as a RIO or Radar Intercept Officer — a backseater. In her squadron she also functions as Assistant Personnel Officer, Public Affairs Officer, Coffee Mess Officer, Educational Service Officer and Morale, Welfare and Recreation Officer. She has been a RIO in F-14s since May 1998 and has, as of November 1999, accumulated about 300 hours in the airplane. "Basically, I help the pilot in his duties. I help with navigation and I'm in charge of the radar and the weapons systems. If the mission we are on is bombs, I set up all the stations for the bombing, I work the radar and back up the pilot in his work. We will make this cruise [six months in the Persian Gulf beginning in January 2000] in the Alphas [F-14A] and probably the cruise after that, and then our squadron will be decommissioned or it will be changed into a Super Hornet squadron."

Dating from the 1950s, the original Dassault-Breguet Etendard transonic strike fighter was planned to be a land-based strike plane for NATO and the French Air

Below: The high-speed F8F Bearcat was designed by Grumman to be Japan's worst nightmare in the final year of World War II. The war ended before the Bearcat could be unleashed against the Japanese, but there was no doubt that the combination of the huge engine in the smallest of airframes would have meant many unpleasant encounters had they been made to face this ultimate cat.

Force, but in the end it was the French Navy that maintained an interest and took it on board for carrier operations. Eventually the plane went through further development and a more powerful, high-performance version emerged. The Super Etendard is, in fact, quite different from the standard model, with a modified high-performance wing, as well as significantly improved avionics and armament. The Super entered fleet service in 1978, with the French Navy accepting a total of seventy-one aircraft. Five were leased to Iraq, while fourteen went to Argentina where they were operated against the British in the 1982 Falkland Islands campaign, from the Argentine carrier, *Veinticinco de Mayo* (originally a *Colossus*-class carrier of the Royal Navy). It was an Argentine Super Etendard naval strike fighter which launched a French Exocet missile to sink the British destroyer HMS *Sheffield* in the Falklands campaign.

For many years Westland of Yeovil, England built a

licensed version of the Sikorsky S-61/SH-3 Sea King helicopter for the Royal Navy. Used in anti-submarine, airborne early warning, and search-and-rescue roles, the RN Sea King HAS Mk 6 and the Sea King AEW Mk 2 are currently operated from HMS *Illustrious* by 820 and 849 Naval Air Squadrons respectively. 820 Squadron is made up of 200 officers and ratings, and seven aircraft. In January 1993 820 Squadron was deployed in support of UN forces in Bosnia (Operation *Grapple*), embarking in the Fleet Auxiliary ships *Fort Grange* and *Olwen*. Its role was switched to logistical support, ferrying men and supplies around the Adriatic. It later continued in the UN support role from HMS *Ark Royal*. 820 Squadron's HAS Mk 6 is a rugged, dependable and versatile maritime helicopter, fully built in the UK and incorporating British electronics and systems. It is able to operate completely independently by day or night and in all weather conditions, making it outstandingly adaptable to search-and-

rescue missions. The normal crew includes two pilots, one observer/navigator/tactical director and one winchman/sonics operator. With a 500-mile range and a four-hour endurance, the HAS Mk 6 is armed with homing torpedoes, depth charges, and a 7.62mm machine gun. Its normal roles are anti-submarine warfare (primary), surface surveillance, search-and-rescue, heavy load lifting, passenger transport and stores delivery, and casualty evacuation.

849 Squadron is the Royal Navy's only Airborne Early Warning squadron, its 'B' Flight consisting of three Sea King AEW Mk 2 helicopters embarked in *Illustrious*. In addition to the AEW role, 849 Sea Kings work with the Sea Harriers of No. 801 Squadron from *Illustrious*, as well as other NATO aircraft, in co-ordinating attacks against enemy aircraft. Their Searchwater maritime surveillance radar can also be used in long-range detection of enemy surface vessels. Over-The-Horizon-Targeting (OTHT) can be employed to guide

attack aircraft or long-range surface-to-surface missiles onto a target. Operated by a pilot, and two observers acting as controller and operator, the 849 Sea King works well in the day search-and-rescue role. In a search for low-flying enemy aircraft, it operates at between 1000 and 6000 feet, depending on atmospheric and sea conditions, and is selected to provide an optimum radar picture. In anti-surface tasks the target is much larger and the aircraft can fly at its airframe ceiling of 10,000 feet. Although Searchwater's maximum range is not fully exploited at this height, ranges well in excess of 100 miles can nevertheless be achieved.

In the 1960s the British Navy and Air Force were told by the Air Ministry to study the possible commonality of their requirements regarding a new aircraft type; a vertical short take-off and landing fighter to be based on the Hawker P1127 prototype VSTOL. The Navy wanted an air superiority, catapult-launched fighter,

Problem: Dead bugs on windshield.
Solution: Live bugs on order.

Problem: Friction locks cause throttle levers to stick.
Solution: That's what they're there for.

Below: The F9F Grumman Panther was the primary fighter of the US Navy in the Korean War. This F9F is landing on the carrier USS *Leyte* (CVA–32) on 6 August 1951.

Above: US Naval aviator and later, test pilot, Bob Elder. Late in 1944, Elder conducted highly-classified carrier-suitability tests with the P–51D Mustang. In his opinion, prior to the Allied capture of Iwo Jima and Okinawa, the US military was concerned about the impending need for long-range fighter escort for the B-29s that would be conducting the final bombing campaign against the home islands of Japan. The ultra-long-range P-51 promised to answer that requirement and Elder set out to see if the "Seahorse" would make the grade. One of his test flights is portrayed in a painting (right) by Craig Kodera and Mike Machat. The P-51 carrier-test project was ended with the American capture of Iwo Jima and its availability as an airfield.

while the Air Force was more interested in a supersonic ground-attack plane. Nearly every time the British and American services are asked by those on high to put aside their differences and find a way to make the same aircraft work for them both, the result is months or even years of pulling and pushing until the concept is finally declared unacceptable by both branches. It happened just that way with this VSTOL attempt. The Navy opted instead to buy the McDonnell F4K Phantom, while the Air Force went on with its feasibility studies until 1965, when the British government cancelled further work on a P1154 supersonic VSTOL variant. By special agreement the government, together with the US and Germany, continued development on the P1127 version which evolved into the Pegasus 3-powered Kestrel. In early tests aboard HMS *Ark Royal* at Lyme Bay in 1963, the Royal Navy showed but minor interest in what it regarded then as little more than a novelty. By 1965 the RAF requirement for such a front line VSTOL plane, primarily in Germany, finally meshed with the ongoing development of the Kestrel. It had been almost totally redesigned to incorporate the avionics of the cancelled P1154, and utilized an uprated Pegasus engine of 19,000 pounds thrust. It entered service with the RAF in 1969 as the Harrier. By April 1973 a new class of ship, a sort of combined anti-submarine and VSTOL carrier, was at last agreed and the first of the class, HMS *Invincible*, was ordered on 17 April. In 1971 No. 1 Squadron RAF had deployed to HMS *Eagle* in an exercise to test their Harrier GR1s in "emergency deployment" to RN helicopter carriers. The exercise was of more than casual interest to the Royal Navy at that point, and they immediately began pursuit of a navalized version: the Sea Harrier. This variant required many major changes, not least being the employment of an uprated 21,500 pound thrust Pegasus 104 engine.

In the 1982 Falklands War, the Sea Harrier, armed with the American AIM-9L Sidewinder missile, provided a wake-up call that the Argentine Air Force, who referred to the plane as "the black death", could not ignore. No Sea Harriers were lost in air combat there, though several were destroyed by enemy action or were casualties of the terrible South Atlantic weather. In 166 days at sea during the conflict, the Sea Harriers of HMS *Invincible* had a 95 per cent serviceability rate. The airplane certainly proved itself, along with the Fleet Air Arm case for the continued operation of fixed-wing aircraft from its carriers.

The Grumman A-6 Intruder joined the US fleet in February 1963 and served there with distinction until the late 1990s. The Intruder stemmed from a Navy requirement in the late 1950s for a carrier-borne day-night medium attack jet capable of delivering a range of weaponry, including nuclear. It was to be a very sophisticated plane with amazing electronic capability and great flexibility of function. The Intruder could carry a massive 18,000 pound load of ordnance on five external store stations under the wing and fuselage, which ranged from conventional iron bombs through such smart weapons as the HARM and Harpoon, and the AIM-9 Sidewinder for self-defence. It carried a radar package enabling it to perform virtually "blind" at low level. Operated by a crew of two, the pilot and the bombardier/navigator sat side-by-side. In the Vietnam war the A-6 often flew in company with other US Navy aircraft, such as the A-4 Skyhawk, and provided bomb-release commands to these less sophisticated aircraft.

Both US Navy and US Marine front line units flew the A-6 for more than three decades, including considerable duty in South-East Asia. After a period of 'de-bugging' common to most new, and nearly all complex aircraft, the A-6A became fully operational and demonstrated that it was more than capable of effectively replacing the very best propeller-driven medium bomber to date, the Douglas A-1 Skyraider. The Intruder was formidable and could operate independently in the worst

Hook Down, Turning Final © 1999, Craig Kodera/Mike Machat.
In the personal collection of Casey Law.

The F-14 is a very good airplane. When it was introduced it was the premier fighter in the world. That was in 1972 and the airplane was built with 1960s technology. Like cars, aircraft are built for varying uses, and fighters, like sports cars, are also different within their specialty. The F-14 was, and remains, the premier fighter for it's role — interceptor and air-superiority weapons platform. It is not the best fighter for dogfights or in an air-to-ground mode. The F-18 is much better in both respects.

By today's standards the F-14 is a difficult fighter to fly. Fighter pilots now have computer protections to remain within the flight envelope to prevent them from losing control or damaging the aircraft. Today's computers are also capable of employing high-lift devices such as slat and manoeuvring flaps to assist the pilot and increase manoeuvring performance. The F-14 had none of these advantages. It was the pilot's responsibility to cope with the g-loading and flap speeds. Of course, the pilot also had to maintain sight of the enemy, his wingman, and control his weapons system with the help of his RIO. Fighting the F-14 is a difficult task for a nugget [new pilot] to learn, and a fleet pilot to maintain. The F-14 also had a characteristic that was always burning in the back of the pilot's mind — the flat spin. We were told never to get hung up with the nose higher than 70° and below 100 knots. Should the aircraft stall and an engine quit, the likelihood was that the plane would go into a flat spin, which was unrecoverable.

weather conditions, something it had to do frequently during the Vietnam war. When no other aircraft could do the work, the A-6 was invariably a reliable performer, putting its bombs precisely on target. With greatly improved avionics, it became better and better over the years and, while not quite as versatile as the Skyraider, the Intruder delivered an over-all level of performance without parallel.

It came to the US Navy in 1961, the big, noisy smoke-dirty McDonnell F-4 Phantom, seven years after the St. Louis airplane builder had been commissioned to design the twin-engined strike fighter-interceptor. Its requirement was changed early in the design stage when the Navy decided that the new plane should be purely a missile fighter. Evidently both the service and the builder had become convinced that the days of close-in dogfighting were over for ever, and that there was no need to carry a gun on the Phantom, a view that found little favour among Phantom crews who were to fly it in air combat. Still, it was the finest aircraft of its type for more than ten years . . . the fastest, highest-climbing aircraft then operating.

The two-man crew of the Navy F-4 sat in tandem, with the pilot doing the flying from the front seat while the RIO operated the powerful long-range acquisition radar from the rear cockpit. The method worked well and both crew members appreciated the second pair of eyes on board in the unforgiving aerial combat arena.

Two missile types were carried as fundamental ordnance on the Phantom. The short-range, heat-seeking Sidewinder for use in distances up to two miles, and the radar-guided Sparrow with a range of thirteen miles. When the airplane entered into air combat during the Vietnam war, it became clear that the lack of an internal gun put the crew at a distinct disadvantage. Early in that war US rules of engagement required Phantom crews to make visual identification of every target, perfectly

reasonable in that most aircraft in Vietnamese skies then were American. The practice virtually eliminated the standoff capability of the Sparrow missile, compounding the F-4 crews' disadvantage in not having a gun, but was overcome with the development of a special radar receiver. Eventually, in May 1967, crew complaints about the armament of the plane were rewarded with the arrival in Vietnam of the F-4D model, which accommodated a pod-mounted 20mm Vulcan cannon. The installation was later modified to be fully internal in the F-4E variant. The powerful Gatling gun gave the Phantom the ability to dogfight on more even terms than its pilots and RIOs had ever known.

The Royal Navy's Fleet Air Arm operated the FG Mk 1 Phantom with No. 892 Squadron at RNAS Yeovilton in March 1969. With the cancellation of CVA-01, the new British carrier intended to replace HMS *Ark Royal* by the late 1970s, the Royal Navy requirement for Phantoms was reduced and a number from its order were transferred to the RAF. These were used to equip No. 43 Squadron based at RAF Leuchars near St. Andrews, Scotland. In July 1972 all of the 892 Squadron Phantoms were relocated to Leuchars which became home to all Royal Navy and Royal Air Force FG Mk 1s in Britain. Powered by Rolls-Royce Spey engines, the FG Mk 1 was slightly faster than the US airplane. It was deployed at sea on *Ark Royal*, the only Royal Navy carrier then capable of operating the Phantom.

Another excellent effort from the designers at Grumman is the EA-6 Prowler, a carrier-borne, all-weather electronic warfare plane. The Prowler came into the US Navy inventory in 1971 and, in 2000, is still the Navy's main performer in that role. It came about in the development of the Intruder strike aircraft, when it became clear that a specialized electronics-based support version of the aircraft was needed to help the Intruders get in and out of enemy airspace. To some extent then, the EA-6 evolved from the Intruder, but the ultimate EA-6B variant has become a very special and distinctive aircraft, with an extremely effective electronic counter-measure stand-off jamming capability. In addition to its highly sophisticated EW systems, the Prowler, which is operated by a crew of four, carries HARM missiles. It has a range of 1100 miles.

With World War II at an end, one of Joseph Stalin's prime goals for the Soviet Union was the creation of a powerful blue-water navy as a counter to the surface fleets of the US and Britain. Stalin died in 1953 with his naval ambitions unfulfilled. From his death until the early 1980s, debate raged in the Soviet Union about the role and composition of their fleet, with the "large aircraft carrier" receiving little support until Admiral Pushkin took the lead in the argument for building a force of American-style carriers. Ultimately, Pushkin won the day, and in January 1983 the keel of the first Soviet super carrier was laid down. Launched in December 1985, the *Tbilisi* was to become the *Kuznetsov,* with the collapse of the Soviet Union when massive defence budget cuts caused the forced sale of the uncommissioned second Russian super carrier, the scrapping of the incomplete third ship of the (Orel) class, and the abandonment of the fourth. *Kuznetsov* was commissioned on 21 January 1991 and features a distinctive ski-jump take-off ramp similar to that of the current Royal Navy carriers *Illustrious*, *Invincible* and *Ark Royal*. It does not incorporate catapults for launching aircraft.

Western observers expect that the air wing of the *Kuznetsov* will eventually contain a combination of the Sukhoi Su-27K and the MiG-29K. Like the American F/A-18, it is believed that the Russians intend to employ the MiG-29K as a multi-role strike fighter. Both aircraft are known to have undergone extensive carrier trials aboard *Kuznetsov*. The Su-27K Flanker-B2 has moveable canard foreplanes which provide added lift at low speeds. They greatly improve the Flanker's take-off, approach and landing performance. It also has both folding wings and folding tailplanes, among many other features tailored specifically to modern carrier operation. Its power is furnished by advanced AL-31F turbofans uprated to at least 12 per cent greater thrust than it's former engines. The Mig-29K is essentially a MiG-29M, but with a strengthened undercarriage, foldable wings, an arrestor hook and other items to make it carrier-friendly. It has a significantly greater fuel capacity than earlier versions of the MiG-29, increasing its range and loiter time. It has new engines rated to provide the additional power needed in relatively short-burst situations like the launch and the bolter or go-around. The navalized MiG is a fly-by-wire analog-controlled airplane for improved handling and extended performance. It has large-area tailerons for improved roll rate, and double-slotted trailing edge flaps for better low-speed characteristics and lower approach speeds. Between the Su-27K Flanker and MiG-29K Fulcrum, the Flanker seems to lack the versatility of the MiG, and the MiG appears to be the superior aircraft in terms of carrier-related performance as well as multi-role capability. In late 1999, however, the composition of the *Kuznetsov* air wing included eighteen Sukhoi Su-27K /Su-33 Flanker D air defence fighters, and four Su-25UT Frogfoot ground attack fighters used for deck training. The big carrier also supports fifteen Kamov Ka-27PL Helix anti-submarine warfare helicopters and three Kamov Ka-31 RLD Helix airborne electronic warfare helicopters.

The principal strike fighter of the US Navy in the 1990s and into the new millenium is the Boeing/ McDonnell Douglas FA/18 Hornet. It first flew in 1978 and entered service in 1980. The Hornet is probably the most capable and versatile fighter-bomber in the history of the American navy. In the Iraq and Bosnian conflicts the Hornet has shown its worth in both the air-to-air and air-to-ground

The aircraft performance was also dramatically affected by the weapons load. The pilot had to consider the addition of drop tanks, missiles, etc.
– Frank Furbish, former US Navy F-14 pilot

Currently operating as the primary anti-submarine aircraft for the US Navy, the Lockheed S-3 Viking was designed in 1972 to replace the S-2 Tracker. The Viking can remain airborne for 7.5 hours giving it a range of 2500 miles and a combat radius from the carrier of 460 miles with 5 hours on station. It can go places at 518 m.p.h. or loiter at just 184 m.p.h. and can descend, using speed brakes, from 30,000 feet to sea level in less than two minutes. In addition to its main function, the Viking has been adapted to handle electronic counter-measures support missions, as well as tanker and cargo roles.

missions. It is a twin-engined, light weight, relatively low-cost fighter offering both rapid acceleration and economical performance. Being fairly small and easily accommodated in the limited space of an aircraft carrier, it has a rather small fuel capacity, and its operations generally involve aerial refuelling. The Hornet was designed primarily as a naval plane to operate from aircraft carriers, but in the inventories of the Canadian, Spanish and Australian air arms, it operates from land bases. It is a highly automated machine whose computers significantly reduce the flying workload of the pilot, freeing him or her to fight the enemy. Its electronic offensive and defensive systems make it an extremely potent adversary.

US Marine Corps F/A-18Ds are two-seat versions in which the back-seater participates in the combat role and can also act as a forward air controller for other aircraft. The Hornet has a gun, the M61 20mm cannon, an electric Gatling gun with a 540-round magazine, in addition to the ability to carry a combination of very capable weapons including the AIM-9M all-angle heat-seeking anti-aircraft missile, the AGM-88 anti-radar missile, and the 1000-pound laser-guided bomb. Pilots and maintenance personnel share respect for the F/A-18's reliability and ease of maintenance, despite the complexity of its systems. In the Gulf War and over Bosnia, the Hornet force had an amazing availability rate of better than 90 per cent, reaching a peak of 95 per cent in the Gulf. Probably no other naval fighter has inspired such admiration and high regard from those who fly it.

The Dassault–Breguet Rafale has been around for years as a single-seat advanced fighter technology demonstrator, appearing at the Farnborough and Paris Air Shows. It has been planned as the basis for a twenty-first century multi-role fighter for France's *Armeé de l'Air* and the *Aéronavale*. The plane is highly manoeuvrable and has good low-speed handling characteristics as well as relatively short take-off and landing capability. Rafale entered flight testing in 1986 but is only entering service with the French Navy in the late 1990s as part of the air defence and strike fighter force in the air wing of the new French carrier *Charles de Gaulle*.

In the competition to win the largest aircraft procurement contract in history, Lockheed Martin and Boeing are, in July 2000, locked in a struggle to build a single, multi-role, supersonic, stealthy fighter with bespoke variants for the US Navy, Royal Navy, US Air Force and US Marine Corps. The Boeing entry is designated X-32, and the Lockheed airplane X-35. The prize: a verified order for at least 3000 of the aircraft.

Long before development of the revolutionary

Left: A-6E Intruders of VA-34 over southern France while participating in the D-Day 50th Anniversary ceremonies in June 1994. Below: A Royal Navy FA 2 Sea Harrier over RNAS Yeovilton, Somerset, England, home of the fixed-wing Fleet Air Arm.

The Dassault–Breguet Rafale was intended to test new aircraft and fighter technologies for both the French Air Force and the French Naval Air Force. It is designed to employ a high angle of attack in combat, to be highly manoeuvrable, to have good low-speed characteristics, and to be capable of relatively short take-off and landing performance. Its maximum speed is estimated at above Mach 2. For most of its developmental years, the Rafale has been a technology demonstrator making mainly static appearances at the Paris and Farnborough Air Shows. Its power is furnished by two 17,000-pound thrust General Electric turbofan engines and much of the airplane is made of carbon fibre and allithium, rather than conventional light alloys.

Right: A 1977 photo of an F-4J Phantom of VF-74, a US Navy fighter squadron embarked on the nuclear-powered carrier USS *Nimitz*, CVN-68. The performance of the big McDonnell Douglas plane was admired by those who flew it. With a top speed of roughly 1500 m.p.h., it was an amazing air-superiority fighter once a gun had been installed. in answer to the demands of the pilots then flying it in Vietnam.

British Harrier jump jet, there had been considerable interest among many in various aviation and military communites in the possibilities for short take-off and vertical landing (STOVL) fighters. Across the decades since the 1950s, no one has really had much to show for their efforts in this direction, and the interest seemed to tail off.

The subsonic British Harrier has proven to be an especially effective fighter for both the Royal Navy and the RAF, but after considering it, the US Navy rejected the plane. It was then that British Aerospace and the US firm McDonnell Douglas teamed up to try and interest both the US and British governments in an up-rated version of the Harrier. No funding was forthcoming from the governments to aid in development, but the US Marine Corps decided to fund a demonstration programme with the goal of improving the plane's range and payload. McDonnell Douglas headed the effort and ultimately the Marines got the result they wanted in the AV-8B Harrier.

Since then, work has continued jointly between the US and Britain on the concept of advanced STOVL (ASTOVL), and following the evolution of various industry proposals, the growth of the Anglo-American partnership in the project, and emerging commitments to it by both the Royal Navy and the US Marine Corps, the current Joint Strike Fighter Programme was under way by the late 1980s.

Which team will win the award to build the JSF? What characteristics will the winning design and, ultimately, the resultant airplane have? How many will actually be built? When will the major decisions be made? They are due to be made early in 2001, but they may not come that soon. Many factors bear on the decisions, not least being the continually changing worldwide US military commitments, and the implications of the American presidential election in November 2000. The answers to the ifs, whats and whens about the Joint Strike Fighter should be fascinating.

TARANTO

Right: A Fairey Swordfish torpedo bomber of the Royal Navy Historic Flight based at RNAS Yeovilton, Somerset, England.

ON THE NIGHT of 11/12 November 1940 two waves of British Fleet Air Arm Fairey Swordfish biplanes took off from the newly-operational Royal Navy carrier HMS *Illustrious* to attack warships of the Italian fleet as they lay at anchor in the harbour of Taranto. A major seaport, Taranto lies in the arch of the Italian boot, about 250 miles south-east of Rome. It had been founded in 706 BC as a Spartan Greek trading post called Taras, and was known as Tarantum in the time of the Roman Empire.

The Royal Navy had been attempting to provide cover and protection for Allied convoys of merchant ships from the constant threat of land-based air attack, from U-boats, and from sorties by the capital ships of the Italian Fleet. The convoys were carrying vitally needed supplies to the island of Malta, whose survival was crucial in the effort to defeat German and Italian forces in North Africa in this early phase of World War II.

British Admiral Sir Andrew Cunningham was the Royal Navy Commander-in-Chief in the Mediterranean from 1939 to 1943. To ease the pressure on his forces in their role of shepherding the convoys, Cunningham ordered a two-carrier strike in the autumn of 1940 on key Italian warships at the Taranto base. The carriers *Eagle* and *Illustrious* were assigned the mission, but various mishaps and a problem with the fuel system of *Eagle* caused her to be withdrawn from the task. Before her withdrawl, five of her Swordfish aircraft and eight air crews were transferred to *Illustrious* for the raid.

A reconnaissance flight by an RAF aircraft on 11 November identified six battleships of the Italian fleet in harbour at Taranto, and the raid was scheduled for that evening. In the late afternoon *Illustrious*, together with her escort of four cruisers and four destroyers, was steaming towards the launch point for the mission. The first wave of twelve Swordfish, fitted with internal auxiliary fuel tanks to augment their range, was being armed. Six of them were to carry torpedoes, four carried

58

Right: HMS *Eagle* whose faulty fuelling system prevented her from participating directly in the attack on the Italian warships in the harbour at Taranto on the night of 11–12 November 1940.
Below: A Royal Navy Swordfish crew with their aircraft.

bombs, and two were taking flares and bombs. The striking force would have to contend with some twenty-two Italian anti-aircraft gun positions, as well as barrage balloons and torpedo nets guarding the harbour.

The British plan for the operation called for the torpedo-armed Swordfish to launch their missiles at the battleships at anchor in the outer harbour. Meanwhile, in a synchronized diversion, the other aircraft would attack cruisers and destroyers moored along the harbour quay. It was thought that the small-warhead 18-inch British aircraft torpedoes could be launched at low level into the shallow waters of the harbour to good effect against the Italian capital ships. It was hoped that the bomb and flare-dropping Swordfish would distract the local defences while the torpedo planes utilized their ultra-low 75-knot cruising speed to weave around and between the barrage balloon cables towards their targets.

The strike force of British aircraft formed up several miles from *Illustrious* and proceeded along the 170-mile route to Taranto harbour by 8.57 p.m., with Lieutenant Commader K. Williamson in command. At 9.15 p.m., however, flying conditions had deteriorated and the squadron became separated in heavy cloud. The planes were compelled to make their attacks independently.

At 10.56 p.m. the aircraft carrying flares and bombs laid an illuminating pattern of flares along the eastern edge of the harbour. They then went on to dive-bomb a nearby oil tank farm which they set afire. At about this time three of the torpedo Swordfish, with Lieutenant Commander Williamson in the lead, had positioned themselves for a diving run on the battleship *Cavour*. They descended from 4000 feet to 400 feet where they released their torpedoes. One of the three torpedoes struck *Cavour*, causing such substantial damage that the ship was never to be repaired. Return fire from the battleship hit and downed

Williamson's plane, but he and his observer survived to become prisoners of war.

Now the second lot of the first wave of Swordfish arrived led by Lieutenant N. Kemp, and their torpedo aircraft attacked the battleship *Littorio*. Their bombing aircraft struck at the cruisers and destroyers, and the seaplane base in the harbour, severely damaging one hangar.

At 11.23 p.m. the second wave of Swordfish began taking off from *Illustrious*, led by Lieutenant Commander J.W. Hale. This force included five torpedo aircraft, two bombers and two planes with flares and bombs. The flare and bomber aircraft made additional strikes on the oil storage facility as the five torpedo-carrying Swordfish assumed a line-astern approach to the battleships. By this time the Italians had begun to put up a fearsome box barrage of anti-aircraft fire from their many shore and ship-based gun positions, and Hale's planes had to fly through the worst of the flak. One of the aircraft was shot down.

While the British striking force lost two aircraft in the raid, they managed to score four hits on the battleship *Littorio*, and one each on the battleships *Cavour* and *Duilio*. *Littorio* was left listing heavily, her deck awash. *Duilio* was run aground, and *Cavour* was listing heavily to port with her stern awash. The following day she settled on the bottom of the harbour. Other ships received some damage, including two *Trento*-class cruisers and two fleet auxiliaries. The oil storage facility and the seaplane base were significantly damaged. The raid was considered to be highly successful. The Italians were left with only two operational battleships. Within six months of the attack, one-third of the Fleet Air Arm air crewmen who had flown it were dead . . . killed in other actions.

It has often been claimed by historians and others that the Japanese drew the inspiration for their low-level torpedo attack on battleships of the US fleet at Pearl Harbor on 7 December 1941 from the

Bill Sarra and his observer, Jack Bowker, were flying in the diversionary bombing element of the first strike of Swordfish on the harbour at Taranto in the evening of 11 November 1940. They had been briefed to drop their bombs on the cruisers, destroyers or oil storage tanks in the inner harbour, the Mar Piccolo. Sarra was unable to discern the specified targets and crossed over the dockyard. It was then that he spotted the hangars of the Italian seaplane base just ahead. He released his bombs from an altitude of 500 feet and watched as one made a direct hit on a hangar, with the others blasting the slipways. The anti-aircraft fire was intense, but Sarra and Bowker returned safely to the carrier *Illustrious*. The Royal Navy victory at Taranto that night meant that the balance of capital ship power in the Mediterranean had been shifted in favour of the Allies. The first awards made to participants in the Taranto strike were few indeed. Six months later in a supplementary list, Sarra and Bowker were 'mentioned in despatches', but by then they were both prisoners-of-war.

When a carrier turns into wind to receive her aircraft, it seems to be a lazy, leisurely movement to the approaching pilot, about to land. The bow starts to turn, imperceptibly to begin with, and then more emphatically in a graceful sweeping arc. The stern seems to kick the other way, as though resisting the motion, and then gives way in a rush, causing a mighty wash astern. While the ship is turning, the wind across the flight-deck can be violently antagonistic to a pilot who tries to land before the turn has been completed. Since we were about to run out of petrol, the turn seemed interminable, and aft one half-circuit of the ship I decided to risk the cross-wind, and the violent turning motion and get down before it was too late.
– from *War in a Stringbag* by Commander Charles Lamb, RN

Right: John Wellham in command of No. 815 Squadron at Alexandria in 1942. Far right: Lieutenant Commander Wellham aboard HMS *Illustrious* at Portsmouth, England in 1999.

Taranto raid. In fact, the Pearl Harbor attack had been thoroughly planned before the British raid was conducted. Still, the method and results of the Taranto raid certainly served to influence and motivate the Japanese forces when the time came to execute their attack.

"I was serving in HMS *Eagle*, a very old carrier which had been in the fleet for some years. We came to the Mediterranean at a time when it was obvious that Mussolini was going to come into the war on the side of the Germans. He'd been sitting on the sidelines, waiting to see which side was likely to produce the most glory for him. As the Germans were sweeping through Europe, it was clearly going to be their side that he wanted to join. We, therefore, were sent to join the Mediterranean fleet, which we did but a few weeks before Mussolini came into the war.

"Led by Sir Andrew Cunningham, a wildly enthusiastic chap, we tore around the Med looking for someone with whom to be hostile, and actually only found the Italian fleet at sea once. We attacked them twice in four hours with our Swordfish torpedo bombers. Unfortunately there were only nine of us making the attacks, quite inadequate against a fast-moving fleet. So it was necessary for us to catch them in harbour. Our admiral did his homework and found that attacking enemy ships in harbour had been quite successful right back to 300 BC.

"The RAF had a squadron of Bristol Blenheims in the area which they were using for reconnaissance of the various North African harbours, including Bomba. One day they told us about a submarine depot ship at Bomba with a submarine alongside it. They did a recce for us the next morning and discovered another submarine coming into the harbour, so the three of us went off in our Swordfish and, at Bomba, spotted the large submarine on the surface, obviously recharging its batteries. Our leader put his torpedo into it and

the sub blew up and sank very satisfactorily. The other chap and I went on and as we got closer we found that, not only was there a depot ship and a submarine, there was a destroyer between the two of them. I let my torpedo go towards the depot ship. My colleague dropped his from the other side and it went underneath the submarine and hit the destroyer. I was very excited and was shouting. Things were going up in the air and we discovered that all four ships had sunk, which was confirmed later by aerial reconnaissance. That night the Italians admitted on the radio that they had lost four ships in the harbour. However, they said that the loss was due to an overwhelming force of motor torpedo boats and torpedo bombers which had attacked them during the night. If I had been commanding officer of that

62

base, I would have reported something along those same lines. I certainly wouldn't have been prepared to admit that three elderly biplanes had sunk four of my ships.

"The attack on the ships in harbour at Bomba had confirmed for us, and for Cunningham, that the best way to attack ships was when they were in harbour. He then decided that an attack on the main Italian fleet in their base at Taranto was imperative. This was all very well, but we in *Eagle* had a limited number of aircraft and no long-range fuel tanks to get us there and back. We also needed absolutely up-to-date reconnaissance to do an attack like that. Then the new carrier *Illustrious* came to join us, bringing us the long-range tanks we needed. The Royal Air Force provided a flight of three Martin Maryland bombers for the reconnaissance work, and they were very good for that. They had the speed, the training and the ability, and each day at dawn and dusk they gave us reports on the positions of ships in Taranto. We couldn't have done the job without them.

"The go-ahead was given for the raid because both *Illustrious* and *Eagle* were fully prepared to

You were throwing the aircraft about like a madman, half the time, and every time I tried to look over the side, the slipstream nearly whipped off my goggles! The harbour was blanked out by ack-ack and I had to check with the compass to see which way we were facing!
– from *War in a Stringbag* by Commander Charles Lamb, RN

do it — but then complications arose. *Illustrious* had a fire in the hangar deck which caused a delay. Meanwhile *Eagle*, which had been bombed repeatedly by the aircraft of the Reggia Aeronautica, had developed plumbing problems. When we tried to fuel our aircraft we weren't quite sure what we were fuelling them with, and so *Eagle* was scrubbed from the operation. We went on with the plan, however, and transferred five aircraft and eight crews over to *Illustrious*.

"The attack was called Operation *Judgement* and it was set for 11 November 1940. Diversionary actions involving merchant navy convoys elsewhere in the Med were organized and worked well, disguising the fact that we were going to hit Taranto.

"We went off in two waves, some of us with bombs, some with torpedoes. The Martin recce flights had shown us that the Italian battleships were in the Mar Grande, the largest part of Taranto harbour, while destroyers, cruisers, submarines and various auxiliary vessels were in the inner harbour, Mar Piccolo.

"At Taranto, I had to dodge a barrage balloon and in doing so I was hit by flak which broke the aileron control spar, and I couldn't move the control column, which is very embarrassing in the middle of a dive. So, using brute force and ignorance, I cleared the column enough to get it fully over to the right. While this was going on and I was trying to get the thing to fly properly, I suddenly appreciated that I was diving right into the middle of Taranto City, which was obviously not a good thing. So I hauled the plane out of the dive, found the target and attacked it. But when you drop 2000 pounds off an aircraft of that weight, it rises. There is nothing you can do about it, and it rose into the flak from the battleship I was attacking. I was hit again and got a hole about a metre long by at least a half-metre wide. The Swordfish still flew and we got back 200 miles over the sea with the aircraft in that condition. It was very painful because, to fly straight, I had to keep left rudder on all the time, which is bad for your ankle, but we got home."

— John W.G. Wellham, RN, Fleet Air Arm pilot

Left: Taranto harbour and the Mar Piccolo twenty-four hours after the attack by Swordfish aircraft from HMS *Illustrious*. Fuel oil is escaping from the Italian cruisers *Trento*, *Trieste* and *Bolzano*. Below: The battleship *Conti De Cavour* on the day after the harbour raid.

SPARKED BY her expansionist policy in the late 1930s, the armies of the Empire of Japan moved first into Manchuria and then were fighting on the Chinese mainland in pursuit of the conquest of South-East Asia. They called their plan the Great East Asia Co-Existence Sphere.

By September 1940 the United States reacted to Japan's incursions by imposing an embargo on her which covered war materials, including scrap iron, steel and aviation spirit, as well as by freezing all Japanese assets in the US.

In a country with few natural resources, Japan's military government knew that they would not survive long without new sources of strategic commodities. They had become persuaded that to win in China they must have the rubber, bauxite, tin, and most importantly, the oil of the Dutch East Indies and Malaya. To get their hands on these resources, Japan's leaders knew that they would have to go to war with the British Empire as well as the Dutch government in exile. They concluded that war with the United States would also be inevitable, as the Americans had a significant military presence in the Philippines and were not likely to tolerate such an adventure on the part of the Japanese. To achieve their goal the Japanese militarists planned attacks that would destroy the US bomber bases in the Philippines, as well as the US Pacific Fleet at Pearl Harbor, Hawaii. They believed that these targets would have to be hit and wiped out to clear the way for Japanese forces to take Malaya, the East Indies and, ultimately, China.

Harvard-educated Admiral Isoroku Yamamoto, Commander-in-Chief of the Japanese Combined Fleet, had once served as a naval attaché in Washington. A brilliant tactician, he knew the military strengths of the United States and had no illusions about America as an adversary. He opposed war with the US but came to accept its inevitability and was quick to warn that when the war came it would be absolutely essential "to give

a fatal blow to the enemy fleet at the outset, when it was least expected". He believed that anything less than the total destruction of the American fleet would "awaken a sleeping giant".

In 1921 a book called *Sea Power in the Pacific* by Hector C. Bywater, a British naval authority, was attracting considerable interest, not least in Japan where by 1922 it had become required reading at the Imperial Naval Academy and the Japanese Naval War College. Bywater took the view that the Japanese home islands were fundamentally protected from direct assault by US forces owing to distance and the secondary fuel and supply consumption involved in such an effort. He proposed that the key to American success against Japan (in a Pacific war) would be by means of an island-hopping campaign through the Marianas, to Guam and the Philippines. In 1925 Bywater published a second book called *The Great Pacific War.* In it his premise was that Japan could develop a virtually invulnerable empire by making a surprise attack on the US Pacific Fleet, invading Guam and the Philippines and fortifying its mandate islands. During a 1934 trip to London, Yamamoto met and spent an evening with Hector Bywater in Yamamoto's suite at Grosvenor House. He was fascinated by Bywater's theories and their implications for Japanese and American strategy.

It was Yamamoto who would conceive, plan and direct the breathtaking surprise attack on the capital ships of the US Navy at Pearl Harbor in the early morning of Sunday 7 December 1941. It was clear to him that it might not be possible to achieve the complete elimination of the US Pacific fleet in one blow. He knew that many American warships were stationed on the US west coast, were undergoing refit, or were in transit between the west coast and Hawaii. His master plan called for a vital second and final strike on what would remain of the enemy fleet, to be staged six months after the Pearl Harbor attack. With only two US aircraft carriers, *Enterprise* and *Lexington*, then

PEARL

Left: With the battleship *West Virginia* burning furiously in the foreground, a small boat rescues a US seaman from the water. Inboard of the *West Virginia* is the battleship *Tennessee*, also ablaze, following the Japanese surprise attack on the US base at Pearl Harbor just before 8 a.m., Sunday 7 December 1941.

Below, left to right: Some of the most successful Imperial Japanese Naval aviators of World War II, Sadaaki Akamatsu, Toshio Ota, Kaneyoshi Muto, Toshiaki Honda, Masuaki Endo and Hiroyoshi Nishizawa.

operating in the Pacific (*Saratoga* was in port on the west coast), Yamamoto liked the odds as he was able to field six Japanese carriers for the Pearl Harbor strike.

In May 1941 senior officers of the Imperial Japanese Navy called on their Italian counterparts at the southern port of Taranto where, in a bold raid on 11 November 1940, two waves of Fairey Swordfish torpedo and bombing aircraft from the Royal Navy carrier HMS *Illustrious* had attacked and done major damage to capital ships of the Italian fleet. Meeting on the deck of the crippled battleship *Littorio*, the Japanese were fascinated by the methods and achievement of the British in

putting half of the Italian fleet out of action in one cleverly planned and executed blow. At a cost to them of only two Swordfish, the RN Fleet Air Arm, it seemed, had solved the problem of how to effectively deliver torpedoes against warship targets at anchor in a relatively shallow harbour. While historians often disagree about the extent to which the Japanese, and Yamamoto in particular, were influenced by the techniques and results of the Taranto raid in the planning and execution of their Pearl Harbor attack, it seems certain that they were inspired and highly motivated by the British example. Thus encouraged, Admiral Yamamoto proceeded with his planning of the attack on the

important forward base of the US Pacific Fleet.

In 1940 both Pearl and Taranto harbours were thought by their warship fleet occupants to be safe and proper anchorages. Both adjoined cities, Honolulu and Taranto, with populations of about 200,000, and both ports were well defended and considered powerful threats to the fleet of any opponent. The US facility and presence at Pearl Harbor was a major impediment to Japan's designs on South-East Asia, while the Italian fleet at Taranto threatened the sea lanes that linked Britain with her interests in Gibraltar, Egypt, Singapore and India.

The British carrier planes, in their effort to dispose of the Italian fleet at Taranto, had to approach the port from the Mediterranean without being discovered by multitudes of Italian reconnaissance aircraft, and evade the fire of fifty-four enemy warships and twenty-one shore batteries. This they had to accomplish while avoiding the steel cables of more than fifty barrage balloons, and the attentions of several squadrons of Italian fighter planes. The matter of overcoming these barriers to success at Pearl and Taranto was shared by Yamamoto and British Admiral Sir Andrew Cunningham, respectively, but the course of history compelled Cunningham to act first, in November 1940, and set the historic precedent. For the Japanese Navy to

America is a large, friendly dog in a very small room. Every time it wags its tail, it knocks over a chair.
– Arnold Toynbee

We [Americans] have a great ardor for gain; but we have a deep passion for the rights of man.
– from a speech in New York by Woodrow Wilson, 6 December 1912

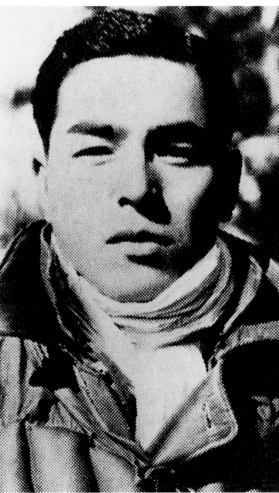

Richard 'Mac' McCutcheon was a powder carman in the number two turret aboard the battleship USS *West Virginia* on 7 December 1941. At just before 8 a.m., dive-bombing and torpedo aircraft of the Imperial Japanese Navy appeared over the battleships of the US Navy that lay at anchor in Pearl Harbor near Honolulu, Hawaii. The *West Virginia* was struck and severely damaged in the attack that followed. Her crew was ordered to abandon ship. Mac McCutcheon jumped into the water and swam for the nearby shore where he found himself but a short distance from the home of a woman who had been witnessing the Japanese aerial attack. She acted quickly when she saw sailors swimming to shore from their burning ships. She gathered clothing from her husband's closet and began distributing it to Mac and the other seamen who had made their way to the shelter of her home.

Right: Among the most famous photographs of World War II is this image of the US Navy destroyer *Shaw*. The ship exploded during the attack on ships of the US Fleet in Pearl Harbor on 7 December 1941. The *Shaw* was not destroyed, however, and was back in service in time to participate in the 26 October 1942 Battle of Santa Cruz.

successfully mount a surprise raid on Pearl Harbor, it would first have to assemble a 32-ship task force, including six aircraft carriers, *Akagi, Kaga, Soryu, Hiryu, Shokaku* and *Zuikaku*, in the Kurile Islands north of Japan, and proceed undetected some 4000 miles, without resupply, to within 200 miles of the Hawaiian Islands. From that point the aircraft of its carriers would have to approach and face the defensive fire of shore batteries, the guns of up to sixty-eight warships, and upwards of a hundred enemy fighter planes.

The British raid on Taranto harbour, the subject of the previous chapter, *Taranto*, is mentioned here as but one factor which, to an extent, influenced the infamous Pearl Harbor attack of a year later.

The shallow harbour at Pearl adjoins the present-day Honolulu airport. The harbour is a basin which surrounds a small airfield facility called Ford Island, and is linked to the sea by only a single narrow channel. The eastern side of the harbour was the site of oil storage tanks, drydocks and a submarine base. On the morning of 7 December, seven battleships of the US fleet, *West Virginia, Tennessee, Oklahoma, Maryland, Arizona, California* and *Nevada*, lay at anchor south-east of Ford Island. An eighth such vessel, *Pennsylvania*, was nearby in drydock.

Probably the greatest challenge faced by the Japanese in their preparations for the Pearl Harbor strike was that of how to successfully deliver torpedoes by aircraft into a shallow basin. The technology of the time, at least prior to the Taranto attack, dictated that a torpedo would virtually always sink to a depth of at least 70 feet before levelling and proceeding toward its target. Such weapons would certainly be hopelessly mired in the mud bottom of the target waters. Impressed by the successes of the British Navy in their raid on the Italian fleet the year before, munitions engineers of the Mitsubishi company near Nagasaki laboured overtime to perfect a missile with a newly-

In the aftermath of the Japanese raid on the American ships and facilities at Pearl Harbor, Hawaii on 7 December 1941, it became clear that the results might have been far more favourable for the Japanese had they been more efficient in the effort.

Certain of the key targets at Pearl, the USN machine shops and the large oil storage tank farm were mainly unaffected by the attack, and most significantly, the Japanese failed to block the narrow and shallow single entrance to the harbour. Had they been able to sink a ship in that tiny channel, they would have denied the US Navy access to Pearl for some time. In failing to destroy the machine shops, they allowed essential repair work to begin immediately on the US warships heavily damaged in the raid. The US ships at Pearl that were still seaworthy, and those that had avoided the attack through being elsewhere, were left able to function and fight owing to the fuel available from the intact tank farm.

In time, nearly 80 per cent of the US aircraft that had been damaged, seemingly beyond repair, at Pearl and other American airfields in Hawaii, were repaired and made fully operational again.

Right: Giving the Imperial Japanese Navy a bit back for the American losses at Pearl Harbor, and in revenge for the loss of the USS *Hornet* (CV-8), aircraft from the USS *Enterprise* (CV-6) delivered the bombs that sank the 13,900-ton Japanese carrier *Zuiho* on 25 October 1944 during the Battle of Leyte Gulf.

designed stabilizer fin. If properly dropped, the redesigned torpedo would sink less than the 40-foot depth of Pearl Harbor and continue on to the target. By 17 November, 180 of the special weapons had been completed and delivered to the Japanese carriers.

Negotiations in Washington with Japan's special envoys there, Saburo Kurusu and Kichisaburo Nomura, to resolve the differences between the two nations by diplomatic means, were clearly failing by 22 November 1941 when, on orders from the Japanese government, the fleet that would attack Pearl Harbor was being assembled. The Japanese believed that the time for talking had run out and their task force headed east from Hitokappu Bay in the Kuriles towards Hawaii on 25 November. The six carriers, escorted by battleships, heavy cruisers, destroyers and submarines, observed strict radio silence as they crossed the western Pacific, in the hope that they would reach the area north of the Hawaiian Islands undiscovered.

On 2 December Task Force Commander Vice Admiral Chiuchi Nagumo received the coded radio signal from Yamamoto, "Niitaka Yama Nabora" ("Climb Mount Niitaka"), authorizing him to commence the attack operation. By 7 December the force had reached their launch position approximately 200 miles north of the island of Oahu and at 6 a.m. the initial wave of planes rose from the carrier *Akagi*. In minutes the aircraft — dive-bombers, horizontal bombers, torpedo bombers and fighters — had launched and were climbing *en route* to their target. Led by Lieutenant Commander Mitsuo Fuchida, the first wave of 183 rounded Barber's Point south-west of Pearl at 7.53 a.m. The calligraphic characters on the white *hachimaki* scarf tied around Fuchida's leather flying helmet translated as "Certain Victory". Using the coded phrase "Tora, tora, tora" ("Tiger, tiger, tiger"), he alerted the carrier force that his planes had reached the objective and were commencing their attack.

AIR RAID PEARL HARBOR. THIS IS NOT A DRILL was the radio message eminating from the command centre at Ford Island a few minutes before 8.00 a.m. as the Japanese torpedo bombers began their strikes on the sleeping ships in Pearl. The first to receive torpedoes were a minelayer, a light cruiser and then the battleship *Arizona*, which was literally blown out of the water, her bottom ripped out. *California* and *Oklahoma* each took three torpedoes, ruining them. A fourth torpedo struck *Oklahoma*, capsizing her. *Arizona* was then hit by a bomb which exploded in her forward magazine, ripping her apart. More than a thousand of her crew died.

At Ewa, Wheeler and Hickam airfields, Japanese bombers and fighters appeared while their colleagues were wrecking the US capital ships in Pearl, and began bombing and strafing runs on the parked aircraft. American Marines and soldiers were strafed as they ran from their barracks at the Hickam base. The attack was followed an hour later by one from a second wave of Japanese carrier aircraft, 167 this time, which further pounded the US facilities.

A total of 2403 military and civilian Americans died in the 7 December raids at Pearl and the other US bases on Oahu. Another 1178 were wounded, 292 US planes were destroyed or damaged, and the 8 US battleships in the harbour were sunk or massively damaged. The following afternoon US President Franklin D. Roosevelt addressed a joint session of the Congress in Washington and called for a declaration of war against the Empire of Japan. Similar US declarations against Germany and Italy followed.

Flushed with success, Admiral Nagumo's Fast Carrier Striking Force sailed back to Japan where it was made ready for the action to come in the escalating Pacific war. What occupied Yamamoto's thoughts in the wake of the Pearl attack, however, was not the recent naval triumph, but rather the fact that not a single American aircraft carrier had been harmed in the execution of his plan.

THE TIDE HAD TURNED in the Pacific theatre of World War II by early 1944. Warships of the US Navy were by then moving steadily westward in their island-hopping campaign, pounding Japanese shore defences before landing masses of invading soldiers and marines whose job it was to re-take these islands from the occupying enemy forces. To take some of the strain from the pilots, aircrews and over-worked sailors of the US Fleet fast carriers, the Americans urgently needed the island airfields that the Japanese held. The stage was set for the greatest carrier battle in history, the First Battle of the Philippine Sea.

Admiral Chester Nimitz was Commander-in-Chief of the US Pacific fleet and in June his Task Force 58, under the command of Vice-Admiral Marc Mitscher, was sent to attack the Marianas and Bonin Islands. On 15 June Nimitz's forces struck at Saipan in the Marianas, some 1300 miles east of the Philippines and more than 3000 miles west of Hawaii. Saipan was defended by Lieutenant General Yoshitsugu Saito and his 32,000 soldiers. They were to meet a US force of 127,000 men, two-thirds of them Marines from an assemblage of 535 US Navy ships under the direct command of Vice-Admiral Raymond Spruance.

In raids on the Bonins, US Navy carrier-based aircraft destroyed 300 Japanese planes for the loss of 22 of their own. The way was then clear for American troops to land on the other beaches of the Marianas with the support of eight escort carriers and their 170 aircraft. In the Marianas campaign, the large US carriers involved included the *Wasp, Lexington, Hornet, Yorktown, Essex, Bunker Hill* and *Enterprise* as well as the smaller escort carriers *Cabot, Cowpens, Bataan, Langley, Belleau Wood, Monterey, San Jacinto* and *Princeton*.

In the months preceding the Marianas attacks, the Japanese suffered considerable losses among their experienced aircrew. Training standards for their replacements had fallen and the remaining

AIR STRIKE WORLD WAR II

I was closing fast. Then I heard a very loud noise. Imagine being in a galvanized tin shed when someone throws a big handful of rocks against the outside. That is what it sounds and feels like when your plane is hit by machine gun fire. It certainly got my attention. I looked back and 300 yards behind me was a pair of Zeros and a stream of red tracers coming at me from the 7.7mm machine gun in the nose of the lead plane.
– Hamilton McWhorter, former US Navy fighter pilot

Left: F6F Hellcat pilots of the USS *Lexington* (CV-16) air group walking to their planes for an air strike against Japanese forces in January 1945.

Below: Alex Vraciu was the fourth highest-scoring US Navy fighter ace of World War II, shown here as the commanding officer of Fighting Squadron 51 in July 1957 which then flew FJ-3 Fury aircraft.

Imperial Japanese Navy aircraft carriers were not at sea much owing to concern about the possibility of their being sunk by US Navy submarines. Japan's carrier force in the Marianas included *Hiyo, Shokaku, Zuikaku, Junyo, Ryuho, Zuiho, Chitose, Chiyoda* and the newly-operational *Taiho*, the largest, fastest and most sophisticated of her carrier fleet. For the Marianas operation the US

Navy carriers had 900 aircraft available to launch, half of which were fighters, with 300 being newly-arrived Grumman F6F Hellcats. For their part, the Japanese Navy had but 430 aircraft ready for action, in addition to some shore-based naval and air force aircraft. The Americans had achieved air supremacy over Saipan. Now Vice-Admiral Spruance sought to deny the use of Iwo Jima to the Japanese as a staging platform for reinforcing aircraft coming from the home islands, by sending half of his carrier task group north to bomb Iwo.

Vice-Admiral Jisaburo Ozawa had brought the Japanese Fleet to the east of Saipan, and on the morning of 19 June, intended to attack the fifteen fast carriers of the recently formed US Fifth Fleet which was under the overall command of Spruance. The American Admiral then ordered his two task group divisions to rendezvous with the balance of his fleet on 18 June, 180 miles west of Tinian. His primary mission was to safeguard the US invasion fleet and the Saipan beachhead, and he chose to maintain a defensive position between Saipan and the approaching Japanese fleet.

Ozawa's second in command, Vice-Admiral Takeo Kurita, planned to hit the American carrier force at a point between his own carriers and the Marianas so his fighters could then refuel and re-arm at Japanese bases on Guam and Rota, and then carry out a second attack on the enemy planes as they returned to the US carriers. Both the Americans and the Japanese had sent submarines into the Philippine Sea prior to the events of 19 June. Reconnaissance planes had pinpointed the position of the US ships for Kurita a day earlier, giving him an advantage over Mitscher, who was still uncertain of the enemy fleet position. Mitscher had to protect his ships and did so by mounting constant combat air patrols over the US carriers.

A series of attacks by shore-based Japanese

planes from Guam began at 10 a.m. and lasted nearly five hours, during which 143 of their aircraft were downed. None of the Japanese planes in these attacks reached the US carriers. In anticipation of further attacks, Admiral Mitscher directed many of his fighters some 50 miles beyond the American carriers where they met and engaged four additional waves of Japanese carrier-based aircraft which suffered grave losses by both US fighter planes and the flak put up from US warships. Of the Japanese attacking force, only 100 aircraft returned to their carriers. Fewer than twenty of their planes got through to the American ships, making hits on the carriers *Wasp* and *Bunker Hill*, causing but minor damage.

Vice-Admiral Ozawa was using the carrier *Taiho* as his flagship on 19 June when it was hit by torpedoes fired from the US submarine *Albacore*, causing fuel leaks. Eight hours later fumes caught fire and a massive explosion ripped the great ship apart. She sank with the loss of approximately 1650 of her 2150-man complement. Four hours after *Taiho* was torpedoed, a similar fate befell *Shokaku*, the victim of another American submarine, the *Cavalla*. After three hours of explosions and spreading fires, *Shokaku* went to the bottom of the Philippine Sea. Vice-Admiral Ozawa had moved his flag to the cruiser *Haguro*.

The next day Ozawa could field but 100 serviceable aircraft. At dusk on 20 June, Mitscher launched 85 fighters to escort 131 strike planes in an effort to destroy what remained of the Japanese carrier fleet. They attacked and sank *Hiyo* and damaged *Chiyoda* and *Zuikaku*, as well as the battleship *Haruna* and the cruiser *Maya*. In the action the Japanese lost sixty-five aircraft against fourteen for the Americans. However, the US pilots faced a 300-mile flight back to their carriers following the air battles of the 20th, and either a ditching at sea when their fuel ran out, or a treacherous deck landing at night. Mitscher, from his flagship, the *Lexington*, elected to illuminate

the flight decks of his carriers to help his pilots through their ordeal, despite the possibility of alerting enemy submarines to the presence of his ships. Still, eighty of the returning US planes were lost when they were forced to ditch or crash, taking the lives of forty-nine American airmen.

The Imperial Japanese Navy never again met the US Navy in real strength during the remainder of World War II. The First Battle of the Philippine Sea was the last major carrier-versus-carrier engagement of the war. Going into the Battle with inferior numbers of warships, aircraft, and aircrews, against a foe that was introducing a new aircraft in the Grumman F6F Hellcat, with superior performance to their own fighters, proved disastrous for the Japanese. Her air groups were wiped out. The loss of three of her aircraft carriers and hundreds of aircraft and aircrew in just two days all but ended her carrier striking capability. Japanese carrier forces would never again be a threat. The one-sided two-day air battle would forever be remembered by US Navy pilots and aircrew as the "Great Marianas Turkey Shoot".

When Alex Vraciu (rhymes with cashew) was an undergraduate student at DePauw University in Indiana in the years just before World War II, he attended a psychology class. During the lecture he shocked the professor and his fellow classmates by suddenly getting up and announcing, "I can't stand this anymore", whereupon he rushed to an open window in the second storey classroom and jumped through it. Coeds screamed, the shaken prof and his pupils went quickly to the windows and gazed down at Alex who was smiling up at them from a tarpaulin being held by several of his fraternity brothers. The leap was a precursor of others. As a naval aviator in the war, he would twice have to bail out of stricken aircraft.

Prior to his graduation from DePauw in 1941, Alex, who like many foresaw the coming war, learned to fly and gained his private pilot's licence

Jack Leaming was a radioman in Douglas SBD dive-bombers during World War II. He was involved in attacks on the Marshall Islands of Roi and Taroa, Wake Island and Marcus Island. In the latter, Leaming and his pilot completed their dive and were immediately bracketed by machine gun fire. Flak then struck an outboard fuel tank setting their plane ablaze. They made a water landing and were taken prisoner by the Japanese. They survived their ordeal and were repatriated in September 1945. "Because there were no landmarks on our long, over-water scouting hops, the pilot had a tremendous navigation task. I always waited for him to initiate conversation . . . didn't want to interfere with his thought processes.

"In dive-bombing, you had to have complete confidence in your pilot and his expertise. In wartime, if he was hit, killed or disabled, and you were still functioning, in all likelihood you were 'going in'. Gravity and speed were against you. There was no time to turn the seat around, put in the auxiliary stick and pull out."

The real issue today is between the frenzy of a herded, whipped-up crowd-begotten cause, and the single man's belief in liberty of mind and spirit, and his willingness to sacrifice his comforts and his earnings for its sake.
– Archibald MacLeish, in World War II

We were shooting three types of rounds: the incendiary, the tracer, and the armour-piercing. The standard belting was 1-1-1, just repeating straight on through. I became convinced that this was not the most effective way to belt them. The armour-piercing would shoot through anything in the Zero. It might have been the preferable round if we were strafing barges with diesel engines in them, but it was an overkill on the Zero, totally unncesessary. The tracer burned out and was pretty much just an empty shell. But the incendiary got our kills for us; there was no question about it after shooting into drums filled with gasoline. The incendiary had enough impact to knock cylinders off of engines (the Zero had no armour plate at that time) and to shoot through anything. Probably nine out of ten Zeros shot down burned. It was very easy to burn the Zero.
– John Bolt, former US Marine Corps fighter pilot

Right: The flight deck of the USS *Essex* (CV-9) with SBD-4 Dauntless and F6F-3 Hellcat aircraft being readied for take-off.

under the government-sponsored Civilian Pilot Training (CPT) programme at Muncie, Indiana. He entered the US Navy as a pilot candidate before the Japanese attack on Pearl Harbor, receiving the gold wings of the naval aviator on 24 June 1942. When he reached the fleet he was lucky in two respects: he was to fly the Navy's then premier fighter, the F6F Hellcat, from carriers in the Pacific, and was to serve in Fighting Squadron 6 (VF-6) under the command of Congressional Medal of Honor winner Lieutenant Commander Edward "Butch" O'Hare.

Vraciu scored his first aerial victory against a Japanese Zero near Wake Island on 5 October 1943. By mid-February 1944 he had nine enemy aircraft to his credit, and when VF-6 was returned to the US, he requested and was granted reassignment . . . to another carrier squadron, VF-16 aboard the USS *Lexington*. It was from the *Lexington* that he participated in the "Marianas Turkey Shoot", after which, for a period of four months, he was the Navy's top-ranking ace.

On the morning of 19 June 1944, Lieutenant J.G. Alex Vraciu and his squadron of twelve aircraft were launched to supplement the *Lexington* combat air patrol which was already airborne. Radar had picked up a force of bogies approaching the US Fleet in several large groups. Vraciu was leading the second of three four-plane divisions. As the Hellcats climbed at full military power, he heard the *Lex* fighter director officer (FDO) radio: "Vector 250. Climb to 25,000 feet, pronto!" The protracted climb at full power was proving too much for some of the Hellcats. Vraciu's engine began to throw a film of oil over his windshield, forcing him to ease back slightly on the throttle. However, his division stayed together and two additional Hellcats joined up with them. All were struggling in the climb and soon accepted that the maximum height they would attain that day was 20,000 feet, a fact they duly reported to the FDO. On the way up to their ceiling altitude, Vraciu

noticed his wingman, Brockmeyer, repeatedly pointing towards Alex's wing. He didn't know what Brockmeyer was trying to communicate until much later when he learned that his wings (which folded to take up less space on board the carrier) were not fully locked. The red safety barrel locks were showing.

The aerial engagement that Vraciu's group was heading for was over before they reached the enemy planes. His group was ordered to come back and orbit at 20,000 feet over the task force. When they arrived over the American ships, the FDO directed the Hellcats to a new heading of 265°. There were bogies about 75 miles out. On the way to the intercept, Vraciu's group saw seven more Hellcats converging from the starboard side.

After flying about 25 miles Alex spotted three bogies and began to close on them. He guessed that there were probably more enemy aircraft in the area and soon saw what he estimated to be at least fifty planes 2000 feet below the Hellcats and on their port side. He was excited and was thinking that the situation could easily develop into a fighter pilot's dream. There did not seem to be any top cover escort with the bogies which, by this time, were identifiable as Japanese aircraft. He picked out the nearest in-board straggler, a Judy dive-bomber, and started a run on it.

Alex peripherally sensed the presence of another Hellcat which seemed to be intent upon the same enemy aircraft that he was after. Concerned about the possibility of being blindsided by the other American, he chose to abort his run, roaring under the Japanese formation and taking the opportunity to quickly look them over. He noted that there were Judys, Jills and Zeroes, pulled up and over the enemy assemblage and selected another Judy out on the edge of the formation. The enemy plane was doing a bit of manoeuvring as Vraciu approached it from behind. The Japanese rear gunner was firing at him as he closed in. He returned fire and the Judy erupted

in flame and began to trail a long smoke plume down to the sea.

Pulling back up to the enemy formation, Alex found two more Judys on the loose, came in from their rear and fired, sending one down in flames. Still on the same pass, he slipped the Hellcat into position behind the other Judy and just as quickly sent it down, its rear gunner continuing to fire at him as the enemy plane fell. Three down.

The great mass of turning, twisting US and Japanese aircraft was moving ever closer to the American fleet. By this time the Japanese formation was badly broken up, but many of its aircraft remained on course toward the US ships. Vraciu reported this fact to his carrier. Then another enemy plane broke formation in his view and he slid over into position after it. He had to be quite careful now and get in very close as his windshield had become increasingly oil-smeared and difficult to see through. A single short burst

proved sufficient to set the enemy plane alight and cause it to enter a wildly out-of-control spin.

The air battle was now much closer to the US ships, and the Jills started descending into their torpedo runs. The remaining Judys were just about to peel off on their bombing runs. Alex saw a group of three Judys in trail and headed for the tail ender. At this point he and they were almost over the outer destroyer screen, but still fairly high. Alex noticed a black puff of flak in the sky near him and realized that the American 5-inch guns were firing at the enemy aircraft in defence of the US fleet. He overtook the next nearest Judy, fired the briefest of bursts and saw the plane's engine come apart in pieces. It alternately smoked and burned as it disappeared below.

Now Alex spotted another enemy aircraft, this one just into its diving attack on an American destroyer. He caught up with the plane and was amazed when yet another very short burst from his Hellcat's guns caused the Japanese to explode, seemingly right in his face. He guessed that his bullets must have hit the enemy's bomb. He had seen planes blow up before, but never like that. He was forced to manoeuvre wildly to avoid the hot, scattering debris of the kill.

As he recovered and climbed back up to rejoin some of the other Hellcats, Vraciu observed that the sky was now entirely free of enemy aircraft. He noticed, too, that a 35-mile-long pattern of flaming oil slicks lay on the water behind them. He later discovered that, owing to his oil-smeared windshield and the need to work in really close to the enemy planes, he had actually fired only 360 rounds of ammunition in shooting down six Judys, all in less than eight minutes.

The next day, 20 June, while flying escort for bomber and torpedo planes in the record 300-mile strike against the Japanese fleet on the second day of the First Battle of the Philippine Sea, Alex Vraciu shot down a Zero, his nineteenth and final victory. For his achievements between 12 and 20

June 1944, he was awarded the Navy Cross.

"On 20 June 1944 the pilots and radiomen of VB-10 were on the alert hoping the Japanese fleet would be found before the end of the day. At about 4.30 p.m. word came down to the ready room that the fleet had indeed been located about 250 miles away. We were very worried about such a long mission so late in the day. If we somehow made it back to our carriers, it would mean night landings. All of our pilots were qualified for night landings but had made only a few in recent months. Despite the drawbacks, we were all eager to take another crack at the Jap fleet. We sure weren't heroes, but this was what it was all about. With the climb out and a full bomb load, the SBD just wasn't supposed to make it there and back.

"We were launched at about 5.30. I remember looking into the sun low on the horizon when we took off. This had to be the most important strike for me personally, since we would be returning to the States after the Saipan invasion. I had been on the *Enterprise* off and on since July 1942, first with Air Group Six and later with Air Group Ten, so I knew this would probably be the last crack at the Jap fleet before the invasion of Japan.

"The flight to the Japanese fleet took about two hours. The time was spent checking out our guns, looking for enemy fighters, and operating our radar, hoping to pick up our target. My pilot, Lieutenant Oliver W. Hubbard, was leading the second section in Skipper 'Jig Dog' Ramage's division. Lieutenant Bangs was leading the other division. Hubbard had agreed that a hit was a must. The Skipper sent Bangs and his division to one carrier, and our division took the *Ryuho*. The anti-aircraft fire was very strong but we managed to get into our dive with no problem, with the exception of a half-hearted attack by several Zeros which were promptly taken care of by our fighters. In our dive we became twisted around trying to get on target. We had to settle for a port to starboard run and

"In mid-February 1944 our air group aboard the USS *Essex* was involved in the first strike against the Japanese at Truk. We were expecting fierce opposition, but it did not occur. After destroying the few enemy aircraft there, our main objective was the bombing and strafing of shipping and shore installations.

"While engaged in strafing, I was hit by anti-aircraft fire at about 400 feet and had to ditch in the lagoon. Fortunately, others in my flight remained over me for a while and were able to intercept an enemy destroyer that was coming to pick me up. They set it afire and caused it to beach on the reef.

"When I landed, my F6F Hellcat sank so fast that I was unable to deploy my raft, and was floating about in my life jacket for three hours after my companions were forced to leave me due to fuel shortage. Then I looked to the north and saw an OS2U float plane from the USS *Baltimore*. It had been dispatched to retrieve me. The pilot, Lieutenant Denver Baxter, made a landing in very rough water, and the rescue was complete." This was one of the first rescues of a downed US pilot from the midst of a Japanese stronghold in World War II.
– George Blair, former US Navy fighter pilot, USS *Essex*

Far left: Deck personnel push an SBD Dauntless dive-bomber into take-off position. Left: US carrier pilots are briefed for a Pacific theatre mission in their World War II ready room aboard the USS *Enterprise* (CV-6).

Below: From the World War II logbook of pilot Tom Harris of VF-17 aboard the US carrier *Hornet* (CV-12), showing flight entries for April 1945. On 16 February 1945, Harris engaged and destroyed a Zero over Tokyo. He shot down two additional Zeros over Yokohama Airdrome on 18 March, and became an ace the next day by downing a Zeke and a Tony near Hiroshima. On 6 April he downed a Tojo near Okinawa, and, as shown below, was credited with a Kate and a George in action on 12 April.

only scored a near miss. On our pull-out, we were headed away from our return heading and had to do a 180 back through the entire fleet, skipping over ship after ship. The battleships were firing their main batteries as we went by. I was strafing everything in sight, mainly out of frustration.

"After leaving the fleet, our main thought was finding our way back to our Task Force. Again, it was about a two-hour flight. Our new radar worked fine and we picked up the *Enterprise* at about 100 miles. There were many frantic calls about fuel problems and going in the drink, but the old faithful SBDs just kept chugging along. As I recall, Jig Dog made only one transmission . . . 'Land on any base.' We did just that, finding the *Wasp* right away. We had barely left the plane and entered the island hatch when a Hellcat crashed the barrier, putting the *Wasp* out of operation for about twenty-five minutes. The next morning I measured our fuel. We had less than three gallons remaining.

Without the excellent landing by Lieutenant Hubbard, we would have taken a swim that night. A flight of over five-and-a-half hours in a plane that wasn't supposed to fly more than about four hours. That's fuel management."
– Jack Glass, formerly with Air Groups Six and Ten, USS *Enterprise* (CV-6)

When Larry Cauble left the west coast of the US aboard the jeep aircraft carrier USS *Barnes* in February 1944, he could not have imagined the adventure that awaited him in the south-western Pacific. "The jeep carrier was built on a merchant ship hull and was smaller and slower than the fleet carriers. It was built utilizing the Kaiser shipyard assembly-line method, which took only about one-fifth of the time needed to build a fleet carrier. The ship was packed stem to stern and the hangar deck was full of aircraft being transported to the combat area to replace losses. There were about

April 45   VF-17   CV-12

Date	Type of Machine	Number of Machine	Duration of Flight	Character of Flight	Pilot	PASSENGERS	REMARKS
1	F6F-5	72901	3.4	STRIKE	Self. Okinawa	102 CL	
2	"	71742	3.3	T-CAP	Okinawa	103 CL	Squadron shot down
2	"	72901	4.1	CAP		104 CL	total of 46 Jap planes
3	"	58072	3.7	SWEEP	Iriomote Shima	105 CL	on 6 April.
5	"	71781	4.5	STRIKE	Kikai Shima	106 CL	
6	"	70546	4.1	T-CAP	Okinawa	107 CL	
6	"	71603	4.0	CAP	"	108 CL	✹ TOJO
7	"	77683	4.3	STRIKE	Okinawa	109 CL	LOST G.W. McADOO – later found in POW camp.
9	"	72896	3.5	CAP	"	110 CL	On 7 April all available planes attacked Jap task force and sunk the BB YAMATO
10	"	77680	4.1	SWEEP	Kikai – Ammamio Shima	111 CL	
12	"	72274	4.0	T-CAP	Kikai	112 CL	✹ Kate ✹ George
13	"	70082	3.3	CAP	"	113 CL	Squadron shot down 3:
14	"	70546	3.3	CAP	"	114 CL	Jap planes on 13 April.
14	"	72257	4.0	CAP	"	115 CL	19 Japs splashed on 13 April
15	"	70546	5.0	CAP	"	116 CL	Japs dropped flares over
16	"	71517	4.6	T-CAP	Okinawa	117 CL	Task force about 0400

thirty of us Ensigns going out as replacement pilots to squadrons in action. The number of additional officers made things very crowded on a ship of that size. We were travelling through the tropics and, of course, in those days there was no air-conditioning. Such niceties as sleeping spaces, cool ventilation, and water for showers, were in short supply.

"It can be a little awkward, joining a squadron already in combat and replacing a pilot who has been killed or lost in action. The squadron members may tend to compare you with their missing friend and squadron mate. I have to say though that in both of the squadrons I joined and flew combat with, I was readily accepted and, after a few days of flying, was treated as a fully-fledged member.

"I was assigned to Fighting Squadron Five (VF-5) aboard the USS *Yorktown* (CV-10) when we arrived at Espiritu Santos in the New Hebrides. After a brief familiarization with the squadron, my first combat flight was to be an attack on airfields on the island of Peleliu. It was 30 March 1944, and I was so nervous and keyed-up that I didn't see a lot of what was going on, including where my own .50 calibre rounds were hitting.

"As we circled back after strafing and sinking a small enemy freighter about five miles off the coast, we saw life boats in the water with uniformed soldiers in them. We knew that the Japanese were shipping replacement ground troops down from the home islands, so we attacked the life boats. The .50 calibre machine guns really tear up wooden boats and human bodies. Between my section leader and myself, we were firing twelve .50s. Pieces of life boats and bodies were flying everywhere. What we didn't know when we opened fire was that, in addition to the soldiers, there were women in the boats. What they were doing there, in an active combat area, I had no idea. Although it was an unpleasant and very graphic sight, if I had it to do over, I would not do the mission differently. We had, of course, read of terrible atrocities committed by the Japanese; Nanking, where some

300,000 Chinese civilians were raped and slaughtered, being just one example, and our squadron had lost three pilots the day before this mission, so revenge was undoubtedly a factor too. Our Marines were then having a difficult time in the ground fighting on Palau, and they certainly didn't need to face fresh enemy reinforcements. War is a messy business of killing people and breaking things. There is no nice way to do it. Obviously, I would have preferred that there had been no women in those life boats, but they were there so they were going to get the same treatment as the combat troops in that situation.

"I had only been with the squadron two months when it was ordered to return to the US to be reformed and readied for another combat tour. I persuaded the CO to let me transfer to a training squadron so I would be sent out again as a replacement pilot. I was ordered to VF-1 and flew a brand new F6F Hellcat aboard the *Yorktown*. On my first day aboard, the commanding officer invited the eleven of us who had just reported in, to his stateroom. He told us that he didn't know why we were aboard, and that he didn't need us. He then chastised several officers for having non-regulation belts or other uniform infractions. That was the last time I had any direct contact with him. VF-1 had been out in the combat area for about five months when I joined them, and had been shore-based on the island of Tarawa before being transferred to the *Yorktown*. After our greeting from the skipper I had no desire to stay with VF-1, so when we got back to Hawaii I requested transfer to the replacement squadron at Barber's Point.

"The replacement squadron felt that our squadron had been out in combat for such a long time that we needed a rest, so they sent us to spend two weeks in a mansion right on Waikiki Beach for a little R and R. It was delightful . . . private beach, surf boards, food prepared by a Navy staff, and a hostess who mothered us and chaperoned when young ladies

Some of the qualities of the better fighter pilots were keen vision, an unselfishness which permitted the individual to be a good team player, an inner calmness and a steady hand. All of these things were of critical importance, but the most critical of all was self-confidence. If you could ask all of the pilots in a squadron to name the five best fighter pilots in the unit, they would all identify the same five or six guys, and each would include his own name in that top five. I always thought I was one of the top two or three of the forty-five pilots in our outfit.
– Ed Copeland, former US Navy fighter pilot

We looked forward to taking on fuel oil at sea, as the tanker would come along side, put their hoses across and then the crew would throw fresh fruit to us on the flight deck. Usually it meant mail call as well, and those long-awaited letters would arrive. Once we were in the ready room reading our mail. Dean Straub and John Trott were neighbors in Norfolk, Virginia, and at the time of our departure from the States, Dean's wife was expecting. Each of them sat there with a handful of mail, and John turned to Dean and said, "Oh, by the way, it's a boy."
– Lynn Forshee, former US Navy radio/gunner, USS *Yorktown* (CV-5)

Eyesight and seeing the enemy first, or at least in time to take correct tactical manoeuvres was very important. However, most important is the guts to plough through an enemy or enemies, and fight it out. There are no foxholes to hide in . . . there is no surrendering. I know of no Navy fighter pilot in the war who turned tail and ran. If one did, he would lose his wings and be booted out of the service for cowardice.
– Richard H. May, former US Navy fighter pilot

Right: F6F Hellcat pilots of Fighter Squadron 16 aboard the USS *Lexington* (CV-16) after their successes against Japanese aircraft near Tarawa on 23 November 1943.

were brought out from town.

"After that pleasant interlude I returned to the war, this time with VF-19 aboard the USS *Lexington,* and a great commanding officer. My fellow pilots there were most welcoming. Friendships and cameraderie develop quickly under combat conditions, not just between contemporaries, but between seniors and juniors, because each is dependent on the other in a tactical situation for protection from attack by enemy aircraft, and other mutual assistance.

"I was assigned as wingman to Lieutenant Bruce Williams. Lieutenant J.G. Paul Gartland was the second section leader. I flew many combat flights with this tactical division. The first five flights were combat air patrol over the fleet. These flights were in defence of the task group and we generally flew in 30 to 40 mile circles at 20,000 to 25,000 feet, waiting for the ship's fighter director to send us to intercept incoming enemy attack aircraft.

"On 21 September 1944 I participated in the first attack of the war on Manila. We were part of a fighter sweep sent to get control of the air. It was to be the first time that I had been fired at by another airplane. Tracer shells going by your wing can get your adrenaline going very quickly. Some pilots lose control of one bodily function or another in such circumstances. It is a very frightening experience. Fortunately, another pilot behind our division shot the attacking plane down before its pilot had a chance to correct his aim. Over Manila one of our planes was badly shot up and the pilot had to bail out. We saw his parachute open, but as he was descending he was strafed by an enemy fighter. The bullets appeared to saw him in half. It was a gruesome sight and made us realize that there was no chivalry in this air war. That particular enemy fighter was then shot down by one of our squadron pilots.

"On 14 October I shot down a Zeke, my second kill. I shared it with 'Willy' Williams since we had both fired at it simultaneously. The task force then

Rare colour images of US Navy World War II pilots. Above right: William LaLiberte. Below left: L.C. Leh, Jr. Below right: John "Jimmy" Thach who devised various fighter tactics including the "Thach Weave", in which a fighter flew across the path of another in order to provide greater protection.

WIN THE WAR BY MAKING *more!*

headed south. Being so close to Japan, the Navy was concerned that the enemy could fly replacement aircraft down to reinforce its units on Formosa, and for the next three days I flew combat air patrol over the fleet as we headed for an area to the east of the southern Philippines.

"Our division was sent out on 21 October to fly west across the Philippines and look for targets of opportunity. We found a wooden barge tied up at the island of San Isadore, and there was no anti-aircraft fire so Willy told us to get ready by sections for a strafing run. As Willy and I made our run in from 5000 feet, we were scoring a lot of hits when the barge suddenly erupted in one massive explosion. Willy's plane was slightly ahead and about 50 feet lower than mine, and he caught the full force of the blast from what had obviously been an ammunition barge.

"Willy's canopy was shattered. His face was cut in so many places that blood was getting in his eyes. He thought he was losing his eyesight. We had been at full power for the strafing run and Willy was still at full power and heading west, away from our ship. We were now 250 miles from the *Lexington* and were concerned about having enough fuel to get back to the ship. We had been out about two-and-a-half hours. Leading the second section, Paul radioed for me to get Willy; that he [Paul] was heading back to the carrier. There was no response to my radio calls asking Willy to reduce power and turn back towards the ship. Finally, I caught up with him and could see that there were big chunks of wood protruding from his wings, pieces of flap and tail section were missing, and his canopy was broken.

"We were using fuel at an alarming rate and heading the wrong way. It took three or four minutes and seemed like an eternity until, with the use of hand signals, I was able to get him to give me the lead. I started reducing power to economic cruise and got us turned around and headed towards the *Lex*. We had a two-and-a-half

hour flight ahead of us, so we exchanged hand signals every few minutes as I wanted to keep Willy alert and informed about fuel management, the distance to go and other flight information. Many of his instruments were shattered, so he could not navigate. When we were 30 miles out from the carrier, I contacted them about our situation. They instructed me to land first. They were concerned that Willy might crash-land and foul the deck if he landed first, leaving me no possibility of coming aboard. Willy seemed to understand my hand signals that he was to land last, but when he saw the carrier he headed straight in without flying a pattern, landed and immediately ran out of fuel. I circled once while they pushed him out of the landing area. Then I landed and ran out of fuel. We had been in the air five hours and thirty-six minutes, a long flight for the Hellcat.

"On 28 October *Lexington* was ordered to the fleet anchorage at Majuro in preparation for her return to Hawaii and the US. As we were leaving the combat area, we were told to transfer some of our best airplanes to the USS *Franklin*. I was one of the pilots selected to fly a Hellcat over to the *Franklin*. We were to fly there, land and be returned by breeches buoy transfer to a destroyer, and then transfer back to the *Lexington*. We expected to be back aboard the *Lex* in two to three hours. It was late in the day and, as there was a slight chance that we might not be able to get back before dark, I decided to take a shaving kit and a khaki uniform with me. We no sooner landed aboard *Franklin* when the fleet was attacked by Betty bombers out from Manila. Our transfer was cancelled, of course. The *Lexington* was hit by a kamikaze which did great damage to the island structure. *Lex* was ordered to leave the task force and proceed to the fleet anchorage at Majuro for emergency repairs. There I was on the *Franklin*, with no way to get back to my clothes and possessions, my squadron mates and my ticket back to the States. Then things got worse.

My day began with breakfast, getting into flight gear, a navigation and intelligence briefing. Along with a complement of fighters and dive-bombers, we took off into a good weather day. Lieutenant Prater brought us into the Jap fleet area and the fighter escort kept what few enemy fighters there were at bay as we approached the *Zuikaku*, which was at full speed and fleeing north, trying to get to safety with the rest of the Japanese diversionary force. Prater put us into our standard torpedo plane attack formation, a 'cut-the-pie' attack which, despite intensive manoeuvring by the intended victims, still resulted in some attackers getting hits. I was one of the last in, as my assigned approach was to have been a port beam shot. As *Zuikaku* manoeuvred in an S-shaped path, I was to have a bow shot. After releasing my torpedo at about 1500 yards and an altitude of about 300 feet, I barrelled in and flew down the starboard side of the ship. We flew through falling streams of phosphorus shot at us as we reformed to head back to the *Lexington*. Our group had made several hits on *Zuikaku*, leaving her smoking. Later, another attack finished her off and by later in the afternoon she was no more.
– Don McMillan, former US Navy torpedo bomber pilot

"The next day the *Franklin* was hit by a kamikaze. Many people were killed, there was a lot of fire and the damage was considerable. I had no general quarters station so I joined one of the pilots from VF-13 in manning a fire hose on the flight deck. I knew several of the pilots there from training and replacement squadron days, and they helped me get a place to bunk. Owing to the extensive damage, there were to be no hot meals for the duration of my stay on board.

"*Franklin* survived (just) to return to Majuro for emergency repairs, so she could get back to the Pearl Harbor shipyard for a more complete repair. Luckily, Air Group 19 was being transferred from the *Lexington* to the *Enterprise*, for transportation back to Hawaii and on to the US, and I was able to get back to *Lex* to pack my things and leave the war zone for the last time."

The US Navy carrier *Hornet* (CV-8) was moving towards a large Japanese naval force near the Pacific island of Midway on the evening of 3 June 1942, when the pilots of Torpedo Eight filed into their ready room. Earlier in the day, B-17s flying from Midway had attacked the Japanese ships, setting two afire.

The torpedo squadron pilots in *Hornet* had gathered to receive the plan of attack that would be employed when the two opposing forces finally engaged. In addition, their Skipper, Lieutenant Commander John C. Waldron, handed them mimeographed copies of his own final message to them, prior to what was to be the Battle of Midway. "Just a word to let you know that I feel we are all ready. We have had a very short time to train and we have worked under the most severe difficulties. But we have truly done the best humanly possible. I actually believe that under these conditions we are the best in the world. My greatest hope is that we encounter a favorable tactical situation, but if we don't, and the worst comes to the worst, I want each of us to do his utmost to destroy our enemies.

If there is only one plane left to make a final run-in, I want that man to go in and get a hit. May God be with us all. Good luck, happy landings, and give 'em hell."

Some of the men of Torpedo Eight then wrote letters to loved ones. It was as if they had premonitions of what lay ahead.
From Bill Sawhill:
"Dear Mom, Dad and Mary,
This letter will be mailed only if I do not get through battle which we expect to come off any hour. I am making a request that this be mailed as soon as possible after I fail to return.

"As you know, I am the gunner and radioman in a plane and it is up to me to shoot first and best. I would not have it any different.

"I want you all to know that I am not the least bit worried, and Mom, you can be sure that I have been praying every day and am sure that I will see you 'up there'.

"Perhaps I have not always done the right thing. Only hope that you do not think too badly of my action.

"The best of luck to all of you and be seeing you when life will be much different 'up there'.

"Be sure and remember I am only one of many and you are also one of many.

Love, Bill"

From Pete Creasy, Jr.:
"Honey, if anything happens to me, I wish you would keep on visiting the folks, for they love you just as much as they would if you were one of their own kids. And by all means, don't become an old maid. Find someone else and make a happy home. Don't be worrying about me and I will be trying to write more often."

And, from Commander Waldron:
"Dear Adelaide,
"There is not a bit of news that I can tell you now except that I am well. I have yours and the

I picked a heavy cruiser of the *Mogami* class to strafe, and as I approached the cruiser from about a mile out at 100 feet above the water, it opened up with every AAA gun *and* their main batteries. The entire ship was flashing and smoking from stem to stern. I can attest to the fact that you can see an 8-inch shell coming towards you. They are spinning slowly, leaving a thin trail of smoke, and you have time to move over out of their way, hoping that they don't explode as they pass nearby.
— Hamilton McWhorter, former US Navy fighter pilot

There set out, slowly, for a Different World, / At four, on winter mornings, different legs . . . / You can't break eggs without making an omelette — That's what they tell the eggs.
— *A War*
by Randall Jarrell

Left: The stubby, unlovely little Grumman Wildcat contributed enormously in the service of the US and Royal Navies in World War II, but she was no match for the Zero.

No aircraft in the world was better suited for its job than the F6F Hellcat. The Hellcat was an extremely stable gun platform with few, if any, bad flight characteristics. It was an easy aircraft to fly and had no hidden quirks such as stalling in slow speed turns like the F4U Corsair. It had one of the most reliable and easily maintained radial engines, the R2800, which was exceptionally rugged, and could withstand a great deal of damage and continue to operate, many times getting a pilot home safely when other fighters would have fallen out of the air. The aircraft itself was also tough and able to sustain massive damage from enemy fire and continue flying. Although several World War II fighters were slightly faster than the Hellcat, the difference was not enough to make an appreciable advantage. In air combat the F6F would out-manoeuvre almost any Allied fighter except, of course, the Spitfire. Most Japanese fighters could out-manoeuvre the Grumman. Here again, the difference in manoeuvrability did not overcome other Hellcat advantages.
— H.B. Moranville, former US Navy fighter pilot

Right: A *Lexington* F6F Hellcat of VF-9 is engulfed in flame after a landing accident on 25 February 1945.

children's pictures here with me all the time, and I think of you all most of the time.

"I believe that we will be in battle very soon. I wish we were there today. But, as we are up to the very eve of serious business, I wish to record to you that I am feeling fine. My own morale is excellent and from my continued observation of the squadron, their morale is excellent, also. You may rest assured that I will go in with the expectation of coming back in good shape. If I do not come back, well, you and the little girls can know that this squadron struck for the highest objective in Naval warfare — to sink the enemy.

"I hope this letter will not scare you and, of course, if I have a chance to write another to be mailed at the same time as this, then of course I shall do so and then you will receive each at the same time.

"I love you and the children very dearly and I long to be with you. But, I could not be happy ashore at this time. My place is here with the fight. I could not be happy otherwise. I know you wish me luck and I believe I will have it.

"You know, Adelaide, in this business of the torpedo attack, I acknowledge we must have a break. I believe that I have the experience and enough Sioux in me to profit by and recognize the break when it comes, and it will come.

"I dislike having the censors read a letter from me such as this, however, at this time I felt I must record the thoughts listed in the foregoing.

"God bless you, dear. You are a wonderful wife and mother. Kiss and love the little girls for me and be of good cheer.

Love to all from Daddy and Johnny"

At 01.11 a message was flashed to *Hornet* telling of the attack by four US Navy PBY Catalina patrol bombers on enemy ships to the south-west of Midway. A little over two hours later "General Quarters" was sounded in *Hornet* and the pilots of Torpedo Eight again assembled in their ready

room, but it was a case of "hurry up and wait". The teletype was silent and the airmen reclined in their big leather chairs, some of them dozing off. The delay was due to Admiral Marc Mitscher and the *Hornet* skipper, Captain Mason, being uncertain of the Japanese fleet's exact position. Without that intelligence, they could not know if the US carriers would be positioned close enough to the enemy ships for the American pilots to operate with a safe fuel margin. Secured from General Quarters at 0600 Midway time, the Torpedo Eight pilots adjourned to the wardroom for breakfast. Just as they finished the meal, General Quarters was sounded again, followed by "All pilots report to your ready rooms." Arriving at their ready room, the torpedo plane pilots found the message: MIDWAY BEING ATTACKED BY JAPANESE AIRCRAFT. Ensign George Gay, a TBD-1 Devastator pilot of Torpedo Eight, recalled: "There was a real commotion as we hauled out plotting boards, helmets and goggles, gloves, pistols, hunting knives and all our other gear. We took down the flight information.

"We had only six of our planes on the flight deck as there was no more room. And since we were to be alone, anyhow, we were the last to be launched. The skipper had tried in vain to get us fighter protection. He even tried to get one fighter to go with us, or even get one fighter plane and one of us would fly it even though we had never been up in one, but he could not swing it. The Group Commander and the Captain felt that the SBDs needed more assistance than we did. They had caught hell in the Coral Sea and the torpedo planes had been lucky. However, the torpedo planes had made the hits in the Coral Sea, so the Japs were going to be looking for us.

"The TBD could not climb anywhere near as high as the dive bombers needed to go, and the Group Commander and the fighter boys did not want us at two levels up there.

"Under these conditions, Commander Waldron

reasoned, our best bet was to be right on the water so the Zeros could not get under us. Since it was obvious that we would be late getting away with nine of our planes still to be brought up from the hangar deck for launching, the problem would be overtaking to form a coordinated attack.

"As I went up to the flight deck, I suddenly ducked into a first aid station and got a tourniquet and put it in my pocket. As I got to the edge of the island, I met the skipper coming down from the bridge. 'I'm glad I caught you,' he said. 'I've been trying to convince them the Japs will not be going towards Midway — especially if they find out we are here. The Group Commander is going to take the whole bunch down there. I'm going more to the north and maybe by the time they come north and find them, we can catch up and all go in together. Don't think I'm lost. Just track me so if anything happens to me, the boys can count on you to bring them back.'

"Each of us would be tracking, and the others all knew how to do it, but it would be my job since I was Navigation Officer. That is also the reason why I was the last man in the formation. It gave me more room to navigate instead of flying formation so closely.

"At a little before 0900, the planes started taking off, and it was 0915 when the signal man motioned me forward to the take-off position. He only had me move forward enough so that I could unfold my wings, and so I began my take-off with the tail of my plane still sitting on the No. 3 elevator. I had more than enough room on deck in front of me to take off, and I noticed as I came back by the ship that the planes that were to follow me were being moved further up the deck with the hope of being able to bring the rest of the torpedo planes of Squadron 8 up the elevator from the hangar deck so they could get in take-off position.

"The rest of Torpedo Eight got off after what seemed to be an eternity, then we all joined up and headed away from our fleet.

"After an uneasy and uneventful hour, the skipper's voice broke radio silence: 'There's a fighter on our tail.' What he saw proved to be a Jap scout plane flying at about 1000 feet. It flew on past us, but I knew — and I'm sure the others did — that he had seen us and reported to the Jap Navy that there were carrier planes approaching.

"We had been flying long enough now to find something, and I could almost see the wheels going around in Waldron's head. He did exactly what I had expected. He put the first section into a scouting line. Each of the eight planes was to move out into a line even with the skipper's wing tips. We had never done this before, only talked about it and, in spite of the warnings and dire threats, the fellows got too much distance between the planes.

Below: A Japanese bomb exploding on the flight deck of USS *Enterprise* (CV-6) killed the Navy photographer, Robert F. Read. It happened on 24 August 1942.

I knew immediately that this was wrong, as the planes on each end were nearly out of sight. The basic idea is O.K. so that you can scan more ocean, but this was ridiculous. I thought to myself, 'Oh, no! Those guys will catch hell when we get back.'

"The skipper was upset, and he gave the join-up signal forcefully. We had just gotten together when smoke columns appeared on the horizon. In less time than it takes to tell, it became obvious that the old Indian had taken us as straight to the Japs as we could fly.

"The first capital ship I recognized as a carrier — the *Soryu*. Then I made out the *Kaga* and the *Akagi*. There was another carrier further on, and screening ships all over the damned ocean. The smoke was from what looked like a battleship, and the carriers were landing planes. The small carrier off on the horizon had smoke coming out of her also. My first thought was, 'Oh, Christ! We're late!'

"The skipper gave the signal to spread out to bracket the biggest carrier for an attack, and that was when the Zeros swarmed all over us.

"Since we had flown straight to the enemy fleet, while everyone else was off looking for them, there was no one else for their air cover to worry about.

"Seeing immediately that the Zeros had us cold, the skipper signalled for us to join back together for mutual protection. We had not moved far apart, so we were back together almost immediately. The skipper broke radio silence again: 'WE WILL GO IN. WE WON'T TURN BACK. FORMER STRATEGY CANNOT BE USED. WE WILL ATTACK. GOOD LUCK!'

"I have never questioned the skipper's judgement or decisions. As it turned out, it didn't make any difference anyhow. We had run into a virtual trap, but we still had to do something to disrupt their landing planes, so he took us right in. We had calculated our fuel to be very short, even insufficient to get us back to the *Hornet*, but this was not considered suicidal by any of us. We thought we had a fighting chance, and maybe after we dropped our fish we could make it to Midway.

UNION OF SOVIET
SOCIALIST REPUBLICS

MANCHURIA

KURILE IS.

ATTU   ALEUTIAN IS.

KISKA   DUTCH HARBOR

KOREA

HIROSHIMA

JAPAN

**NORTH PACIFIC
OCEAN**

CHINA

HONSHU

TOKYO

SHIKOKU

NAGASAKI

KYUSHU

IWO JIMA   MARCUS

OKINAWA   MIDWAY

MARIANAS IS.   S

FORMOSA

RMA

CHINA SEA

PHILIPPINE SEA

WAKE   HAWAIIAN IS

PEARL HARBOR

PHILIPPINE IS.

SAIPAN   TINIAN

THAILAND

ROTA   GUAM   ENIWETOK   KWAJELEIN

MAJURO

MANILA   SAMAR   MARSHALL IS.

LEYTE GULF   ULITHI   TRUK

INDOCHINA   YAP

SINGAPORE   CAROLINE IS.

MAKIN   GILBERT IS.

BORNEO   TARAWA

ADMIRALTIES

RABAUL

SOLOMON IS.

JAVA   CELEBES   BOUGANVILLE

NEW GEORGIA

NETHERLANDS   NEW GUINEA   TULAGI   SANTA CRUZ IS. NDS

EAST INDIES   PORT MORESBY   GUADALCANAL

SAVO   SAMOA IS.

HE PACIFIC WAR THEATER
F WORLD WAR II   NEW HEBRIDES

CORAL SEA

AUSTRALIA

NEW CALEDONIA   NOUMEA

Things then started happening really fast.

"I cannot tell you the sequence in which the planes went down. Everything was happening at once, but I was consciously seeing it all. At least one plane blew up, and each would hit the water and seem to disappear.

"Zeros were coming in from all angles and from both sides at once. They would come in from abeam, pass each other just over our heads, and turn around to make another attack. It was evident that they were trying to get our lead planes first. The planes of Torpedo Eight were falling at irregular intervals. Some were on fire and some did a half-roll and crashed on their backs, completely out of control. Machine-gun bullets ripped my armour plate a number of times. As they rose above it, the bullets would go over my shoulder into the instrument panel and through the windshield.

"Waldron was shot down very early. His plane burst into flames, and I saw him stand up to get out of the fire. He put his right leg outside the cockpit, and then hit the water and disappeared. His radioman, Dobbs, didn't have a chance. Good old Dobbs. When we had been leaving Pearl Harbor, Dobbs had orders back to the States to teach radio. But he had chosen to delay that assignment and stay with us.

"Much too early, it seemed, Bob Huntington [Gay's radioman/gunner] said, 'They got me!' 'Are you hurt bad?' I asked. I looked back and Bob was slumped down almost out of sight. 'Can you move?' I asked. He said no more.

"It was while I was looking back at Bob that the plane to my left must have been shot down, because when I looked forward again it was not there. I think that was the only plane I did not actually see get hit.

"We were right on the water at full throttle and wide open which was about 180 knots. Anyone slamming into the sea had no chance of survival — at that speed the water is just like cement. That is why I was so sure that they were all dead.

"It's hard to explain, but I think Bob being put out of action so soon was one of the things that saved me. I no longer had to fly straight and level for him to shoot, so I started dodging. I even pulled up a few times, and took some shots at Zeros as they would go by. I am positive I hit one, knocking Plexiglas out of his canopy. I may have scared him, but I certainly did not hurt him personally, or even damage his plane much.

"The armour-plate bucket seat was another thing that worked well for me. I could feel, as well as hear and sometimes see, those tracer bullets. They would clunk into the airplane or clank against that armoured seat, and I had to exercise considerable control over when to kick the rudder.

"About this time I felt something hit my left arm and felt it to see what it was. There was a hole in my sleeve and I got blood on my hand. I felt closer, and there was a lump under the skin of my arm. I squeezed the lump, just as you would pop a pimple, and a bullet popped out. I remember thinking, 'Well, what do you know — a souvenir.' I was too busy to put it in my pocket. So I put the bullet in my mouth, blood and all, thinking, 'What the hell — it's my blood.'

"We were now in a position, those of us still left, to turn west again to intercept the ship we had chosen to attack, but the Zeros were still intent on not letting us through, and our planes kept falling all around me. We were on the ship's starboard side, or to the right and ahead of our target, and as we closed range the big carrier began to turn towards us. I knew immediately from what the skipper had said so often in his lectures that if she got into a good turn she could not straighten out right away, and I was glad that she had committed herself. At that moment there were only two planes left of our squadron besides my own. One was almost directly ahead of me, but off a bit to my left. I skidded to the left and avoided more 20mm slugs just in time to pull my nose up and fire at another

Zero as he got in front of me. I only had one .30 calibre gun, and although I knew I hit this Zero also, it did little damage. When I turned back to the right, the plane that had been directly ahead of me was gone, and the other one was out of control.

"My target, which I think was the *Kaga*, was now in a hard turn to starboard and I was going towards her forward port quarter. I figured that by the time a torpedo could travel the distance it should be in the water, the ship should be broadside. I aimed about one-quarter of the ship's length ahead of her bow, and reached out with my left hand to pull back the throttle. It had been calculated that we should be at about 80 knots when we dropped these things, so I had to slow down.

"I had just got hold of the throttle, when

Left: Photographed during the Battle of Santa Cruz in October 1942, fighter ace and *Enterprise* pilot Stanley "Swede" Vejtasa of VF-10 shot down seven Japanese aircraft in a single sortie. In later years he became skipper of the *Kitty Hawk*-class aircraft carrier *Constellation* (CV-64). Below: A pilot from the *Enterprise* is interrogated after a World War II sortie from the ship.

Below: An SB2C Helldiver prepares to take off from a World War II *Essex*-class carrier in the Pacific. Far right: Fleet Admiral Chester W. Nimitz was generally credited with masterminding the victorious US campaign in the Pacific. Bottom right: Vice-Admiral Marc A. Mitscher, who commanded US Navy Task Force 58 in the Pacific campaign. Bottom centre: Admiral William F. Halsey, one of the principal US Navy commanders in the Pacific during the war. Below: The 1942 issue of Life magazine featuring an article on George Gay.

something hit the back of my hand and it hurt like hell. My hand didn't seem to be working right, so I had to pull the throttle back mostly with my thumb. You can well imagine that I was not being exactly neat about all this, I was simply trying to do what I had come out to do. When I figured that I had things about as good as I was going to get them, I punched the torpedo release button.

"Nothing happened. 'Damn those tracers', I thought. 'They've goofed up my electrical release and I'm getting inside my range.' I had been told that the ideal drop was 1000 yards range, 80 knots speed, and 80 feet or so of altitude. But by the time I got the control stick between my knees and put my left hand on top of it to fly the plane, and reached across to pull the cable release with my good right hand, I was into about 850 yards. The cable, or mechanical release, came out of the instrument panel on the left side, designed to be pulled with the left hand. But those damn Zeros

had messed up my program. My left hand did not work. It was awkward, and I almost lost control of the plane trying to pull out that cable by the roots. I can't honestly say I got rid of that torpedo. It felt like it. I had never done it before so I couldn't be sure, and with the plane pitching like a bronco, I had to be content with trying my best.

"God but that ship looked big! I remember thinking, 'Why in the hell doesn't the *Hornet* look that big when I'm trying to land on her?'

"I remember that I did not want to fly out over the starboard side and let all those gunners have a chance at me, so I headed out over the stern.

"I thought, 'I could crash into all this and make one great big mess, maybe even get myself a whole carrier, but I'm feeling passably good, and my plane is still flying, so the hell with that — I'll keep going. Maybe I'll get another crack at them and do more damage in the long run.'

"Flying as low as I could, I went between a

couple of cruisers and out past the destroyers. If you have ever seen movies of this sort of thing, you may wonder how anything could get through all that gunfire. I am alive to tell you that it *can* be done. I think my plane was hit a few times . . .

"The Zeros had broken off me when I got into ack-ack, but they had no trouble going around to meet me on the other side. A 20mm cannon slug hit my left rudder pedal just outside my little toe, blew the pedal apart and knocked a hole in the firewall. This set the engine on fire, and it was burning my left leg through that hole.

"When the rudder pedal went, the control wire to the ailerons and the rudder went with it. I [still] had the elevators so I could pull the nose up. Reaching over with my right hand, I cut the switch. That was also on the left side. I was able to hold the nose up and slow down to almost a decent ditching speed.

"Most airplanes will level out if you turn them

loose, especially if they are properly trimmed. Mine was almost making it, but I was crosswind, so the right wing hit first. This slammed me into the water in a cartwheel fashion and banged the hood shut over me before it twisted the frame and jammed the hood tight.

"As I unbuckled, water was rising to my waist. The nose of the plane was down, so I turned around and sat on the instrument panel while trying to get that hood open. It wouldn't budge. When that water got up to my armpits and started lapping at my chin, I got scared — and I mean really scared. I knew the plane would dive as soon as it lost buoyancy and I didn't want to drown in there. I panicked, stood up and busted my way out.

"The Zeros were diving and shooting at me, but my first thought now was of Bob Huntington. I was almost positive he was dead. I think he took at least one of those cannon slugs right in the chest, but I thought that the water might revive him and I had

to try and help him. I got back to him just as the plane took that dive, and I went down with it trying to unbuckle his straps and get him out.

"The beautiful water exploded into a deep red, and I lost sight of everything. What I had seen confirmed my opinion of his condition and I had to let Bob go. The tail took a gentle swipe at me as if to say goodbye and I came up choking. I lost the bullet from my mouth and as I watched it sink in that blue, clear water, I grabbed for it but missed. Zeros were still strafing me and I ducked under a couple of times as those thwacking slugs came close. As I came up for air once, I bumped my head on my life raft out of the plane."

George Gay was the only member of Torpedo Eight to survive their attack that day. With his wits and luck, he evaded capture by the ships of the enemy fleet that drove past him as he floated near them, his head covered by a thin black seat cushion that had emerged from his sinking plane, along with a four-man life raft. That entire afternoon, in the presence of the Japanese warships, he "rode" that uninflated raft like a horse, concealing it and himself from his foes, having to wait until nightfall to inflate it. From his vantage point, he watched the destruction of three great carriers of the enemy fleet by the dive bombers of *Hornet* and other US flattops. After about thirty hours in the water, Gay was saved when he was sighted by a PBY crew who landed nearby to pick him up. He was flown to Midway and later to Pearl Harbor where a doctor examined him and noted that the ensign had lost roughly a pound an hour in body weight during his time in the sea. While recovering, George Gay was visited in his hospital room by Admiral Chester W. Nimitz, Commander-in-Chief of the US Pacific Fleet. The admiral had a more than casual interest in the ensign's account of what he had observed in the course of his amazing adventure. Gay later received the Navy Cross and the Purple Heart.

Far left: Hellcat pilot Wilbur Webb. Left: George Gay of VT-8 with Scott McCuskey of VF-3. Below: Exhausted night crew in their sacks aboard USS *Enterprise* (CV-6), 27 May 1945.

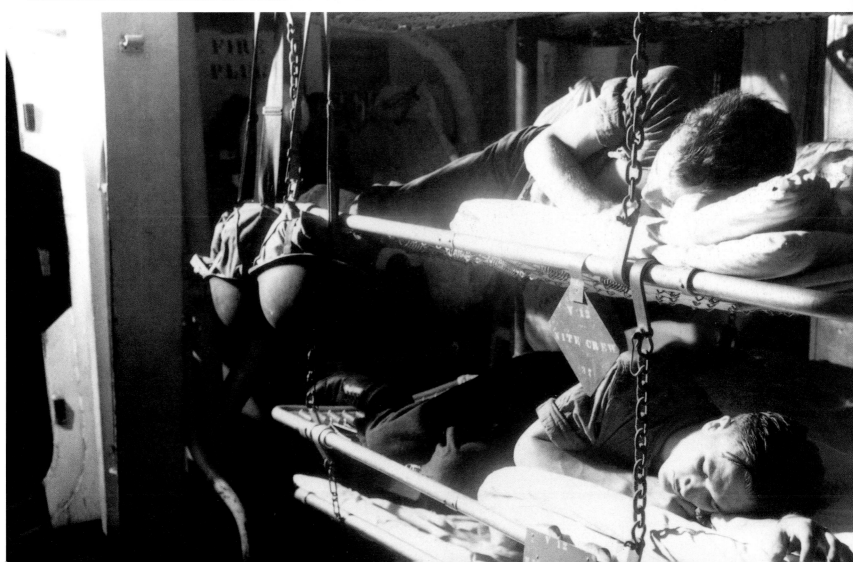

THOMAS LAGRANGE has been on the USS *John C. Stennis* (CVN-74) for a year. As the Carrier Air Traffic Control Chief, his job is to assist the CATC Watch Officer in the proper functioning of CATC: the proper separation and sequencing of aircraft. Today the weather is fine with good visibility and no ceiling. The pilots are essentially handling air traffic control themselves: "We're not doing a whole lot until it gets dark, and then we'll take control of the pattern. They won't just run around and bounce any more. We'll stack them up and give a positive control, but while the weather is good, they do the majority of the work. If the weather goes bad and the ceiling drops below a thousand feet, or the visibility reduces to below five miles, then we'll take over.

"The people we are qualifying now are from our own Air Wing. They are seasoned pilots who just have fallen out of currency, so they are being re-qualified. They come out to the ship and do at least four day and four night landings, and then they fall back within currency. When you get guys that have never qualified before, they come out and we take them from beginning to end, getting them initially qualified.

"We don't stage them through dusk from day to night. There is either day or night. It's one or the other. Even if the weather is below minimums, it's still day or night, so they fly in the day and then they stop. We refuel the jets, they go to chow and then they hop back in after sunset when it's dark and start flying again. When they get airborne, it's dark. We give them fifteen or twenty minutes to get comfortable with the aircraft and then give 'em a shot. They do the approach so they can trap or touch and go, whatever, but it's either day or night.

"Helicopters will always be airborne, on station, for plane-guard rescue operations. They'll be within five miles of the ship at night, and within ten miles in the daytime.

"My job is never monotonous. The pace changes. Just about the time you might start to get bored

AIR TRAFFIC CONTROL

Left: Royal Navy Fleet Air Arm Commander George 'Wings' Wallace heads the Air Department in the *Invincible*-class carrier HMS *Illustrious*.

All movement on the 4.5 acre flight deck of a modern American supercarrier is managed by Flight Deck Control (FDC) which is operated in a room at the base of the island. Using a tabletop layout of the flight deck to scale, the Handler and his assistant spot the aircraft, moving miniature top-view aircraft shapes. Coloured pins, bolts and wingnuts on the shapes are used to indicate aircraft condition such as "needs fuel". FDC orders various deck crew tasks as needed, from fuelling to ordnance loading, to the movement of planes to the catapults. All orders leading to the launch of aircraft eminate from FDC.

Located in the island, high above the flight deck, Primary Flight Control (PriFly) is the office of the carrier Air Boss and his assistant, the Mini Boss. It affords the best perspective of the massive flight deck and area surrounding the ship. Activity in PriFly can be intense during flight operations, with several people performing their functions at once in a relatively small area. The Air Boss and Mini Boss are responsible for control of the air space within ten miles of the carrier.

sitting around, the pace picks up and challenges you for awhile, makes you tired. Then you stop, everybody rests, and then you come back and do it again. It's not a nine-to-five job, it's always different. You never know what you're gonna be doing. Our shift is flight ops to flight ops, so even if it's 1200 to 2400, a twelve-hour day, you show up a little bit early to be prepared, stay a little longer to debrief, so thirteen, fourteen hours. Right now we're getting some help from the weather so it's an easy twelve or thirteen hours. If the weather doesn't co-operate, or we have a man out, then it's twelve or thirteen hours of constant work.

"When the weather is rough and the sea swells are high, all *we* can do is give them a shot at the approach. When they get down on the ball, then we're done with it, so their start on the ball has an impact, but it's all from three-quarters of a mile in where you make your money. You want to give them a start on the ball, but we drop them off at three-quarters of a mile and the LSO takes over. Hopefully, we've got him [the pilot] in a position where he can get a look at the deck.

"Before the *Stennis*, I was on the *Independence*, from 1990 through 1993, including the Gulf War. It was a big test for us because nobody had any confidence in the military. We were over there and everybody thought we were gonna get shellacked. But we were pretty good at [what we had to do there]. We had trained for it all along, and when we got in there we were able to do the job with precision. It was a new era for us.

"I like my job very much. I'd like to see more people in the military, but, with the cutbacks . . . the peacetime Navy . . . it goes with the times. They [the US Congress] figure the threat has diminished so whenever they want to cut money they start reducing the number of [military] personnel which, in the long run, has an effect. It's hard to keep people [in the Navy]. The civilian sector is gonna pay someone [an air traffic controller] fifty or sixty thousand dollars to do

what they are getting paid chicken-feed for doing in the military now, so you can't hold it [leaving the service] against them. You can't begrudge somebody if they have the gumption to go out and make it on their own. There is good and bad out there. You pick your own destiny. There is either a terrific dedication factor [in the military] or a dependency factor. People who are dedicated, they want to do well for themselves, and this is something they want to finish. You've also people who are making enough money to suit their needs and have become dependent on the steadyness of the income.

"There is a lot of stress and pressure. Now though, we have e-mail and we've got sailor [satellite] phones, so we can call home. It's a little expensive but the capability is there to call home. That's a good thing. So is e-mail. You get same-day response, whereas if you write a letter you've got to wait two or three weeks for a response and that tends to add a little pressure on you."

Royal Navy Fleet Air Arm Commander George Wallace, Commander (Air), is head of the Air Department, responsible for all aspects of flying on board HMS *Illustrious*. The occupant of his job on a Royal Navy carrier is known informally as "Wings", and is equivalent to the "Air Boss" on a US Navy carrier. The Air Department in *Illustrious* is in two parts. One is the ship's company, who are with the ship permanently, and the other is the squadrons, which are not always with the ship. When the squadrons are embarked, they fall within the Air Department. When they are disembarked, which can be for months at a time, they go back to their air stations, such as RNAS Yeovilton in Somerset. While they are disembarked it is the responsibility of the air stations to make sure that they deliver the squadrons back to the ship, operationally capable and ready to do their jobs.

"I talk with the air stations so that they have a good understanding of the requirement here.

When the guys come [to the ship] with their airplanes, they have been properly trained and are fit for purpose. We then get them on board and go off and do what we need to do.

"Normally we have on board a squadron of Sea Harriers (six or seven aircraft), a squadron of anti-submarine warfare Sea King helicopters (seven aircraft), and a flight of electronic warfare Sea Kings. Recently we had a flight of RAF Harrier GR7s. These have embarked from time to time for training and for operations in the Gulf and the Adriatic in recent years. That is the way of the future, a joint package of capability which can be delivered from the sea, to combine the fighter sweep, escort, and combat air patrol, with the GR7's air-to-ground capability. It is very much today's business and the business of the future.

"I think the RAF came to the ship with a bit of trepidation, not quite knowing what they were letting themselves in for, but it worked very well and they went away thinking 'we can do this'. There is a lot of work needed before embarkation to make sure that any difficulties or problems, either from our perspective or theirs, are ironed out before they get here so they are comfortable with coming on board. We need to work quite carefully together so that when we deploy that capability for real, it's not something that's new. It's something we're all comfortable with, that we have worked towards and that is going to be a continuous process. It is carefully linked into Joint Force 2000, which will be a coherent organization putting RAF Harriers and Navy Harriers together with Nimrods and search-and-rescue Sea Kings. For the ship it will be made manifest in joint carrier operations, and with future carriers and future carrier-borne aircraft.

"Joint Force 2000 will enable a joint view towards how we are going to deliver capability at sea and from ashore. Putting the Harrier force together, and that's going to happen quite soon, is the first step in a joint expeditionary-style operation.

105

Top: An F/A 18 is directed onto the number one catapult. Bottom: The Air Officer in PriFly aboard the USS *Ranger* (CVA-61) in 1959. Right: An F/A-18F Super Hornet launches from the USS *John C. Stennis* (CVN-74) in 1997.

"Most RAF guys didn't join that service to go to sea. The crucial thing, though, is to be able to deliver the capability I'm talking about. In my view, that means a properly planned sequence of training events to build that capability and maintain it. The special needs of operating aircraft at sea are such that you can't just expect to have them fly aboard and, five minutes later, it all works. It doesn't. It needs a lot of hard work to bring it all together and an appreciation that this [policy] has been developing over the past few years. With that appreciation, we know the size and shape of the task and can apply the right sort of effort to be sure that the end product is useful and efficient, and delivers what the government wants.

"We have to fly at a certain level or standard to maintain the skills, and practice them daily, weekly or whatever. The ship's programme is really the driver in putting together a programme of activity to maintain or work up skills for a particular operation.

"My job falls into two main parts. One is planning what we are going to be doing, and the other is doing it, on a day-by-day or week-by-week basis.

"The recent integration of the RAF GR7s brought with it a degree of uncertainty about what was going to happen. These were guys who hadn't flown to the deck before, so that was an interesting, heartbeat raising period. When we are doing an exercise, it gets really busy. For example, using the Harrier FA2s to generate a two-over-two combat air patrol on a continuous basis . . . that's quite a drain. On the fixed-wing side alone you end up launching and recovering aircraft every hour and a quarter throughout the day. There are peaks of activity punctuating the day, and on top of all that you have the helicopter flying, which is going on throughout. All that is quite challenging, but we have watchkeepers in the Air Department so we can keep that sort of thing going 24 hours a day. The thing I have difficulty with is there is only

one of me, and keeping it going 24 hours a day relies on my being around when the plans are coming together, and when potentially tricky events are happening. That's quite tiring. Generally in an exercise, I am up and working from five in the morning until midnight the next day. Once you get into that sort of routine, it's OK, but making the transition from a more normal working regime to that regime takes a few days. My principal job, in terms of the day-to-day conduct of flying, is maintaining the wide picture so we don't forget things . . . so that the things that happen at short notice can be picked up while all the other things are going on all around us."

Understanding the nuances of fixed- and rotary-wing aircraft can be tricky. Commander Wallace comes from a helicopter background and he understands the Sea King. His last job, prior to *Illustrious*, was at RNAS Yeovilton where he was responsible for delivering the Sea Harrier jet "product" to the front line, a job he held for two-and-a-half years. "I have that much understanding of the Sea Harrier force, but the Sea Harrier is a single-seat fighter and unless you're a single-seat fighter pilot you never really know what it's like. I have a feeling and interpretation of what it's like, but in this job I have to rely heavily on the fixed-wing expertise of our Harrier pilots, the guys who *do* know. You can't, of course, be skilled in every aspect of flying, and it's the mix of the team here that's important.

"Certainly, the fixed-wing stuff is the most interesting flying we conduct. Generally, the helicopters go about their business and, once we have the plan together, that's fine. With the fixed-wing side, there are always going to be peaks of activity. The perceived or actual pressure to get the jets launched on time to meet their operational target . . . if it's during an exercise, there is not quite so much pressure. In an operation though, where you've got, say, four FA2 Sea Harriers and four GR7s launching at the same time to go rendezvous with a much bigger package at a particular time, getting them off the deck is quite a challenge. If it is at night or in dodgy weather, or both, it is more of a challenge. The most difficult and stressful time is when you have everything on the margins: marginal weather, airplanes with marginal performance and marginal fuel states, all coming back at the same time, and you have to get them on the deck because there is nowhere else for

them to go. That is, perhaps, the most testing of times, but it can be done and *is* done regularly. The more we practise it the more comfortable we are that it can be done, and done safely and efficiently – and getting things done safely is what it is all about. Whether in peacetime or in support of an operational mission, it's got to be done safely.

"The Sea Harrier is a very capable airplane, but it is a single-seat fighter and the pilot has got a lot of

Below: An F9F Panther takes the wave-off from the landing signal officer of a US carrier during the Korean War in the early 1950s.

Three key developments in the evolution of the aircraft carrier are British in origin. The first of these was the angled deck, whose trials in the early 1950s proved both successful and inspiring. While the Royal and US Navies remained skeptical, they did convert one carrier each, *Triumph* and *Midway* respectively, to be testbeds for the idea. Aircraft carrier flight decks prior to this had been designed with an axial deck. The front one-third of the deck was the province of aircraft awaiting their next take-offs. The balance of the deck to the rear was dedicated to recovering aircraft. A landing pilot generally had few options. Either he snared one of the arresting wires with his tailhook, or he continued down the deck and into a barrier or the parked aircraft at the sharp end.

The concept of an angled flight deck area occurred to engineers of the Royal Aircraft Establishment at Farnborough. They believed that, utilizing a deck that offered a runway angling to port from the stern, a pilot whose plane failed to catch an arresting wire would be able to apply power and go around for another approach. By mid-1953, more than 4000 landings were successfully achieved in trials of the angled deck on the USS *Antietam* (CV-36). Following the trials, other US and British carriers were refitted with angled flight decks.

Next came the steam catapult. New efficiency in flight deck

work to do, whether it is day or night. At night his visual cues are constrained, so it is all the more difficult. As for the pilot's psychological state, anything we can do to ensure that his job is as straight forward and easy as it can be, we should do. On recovery, the wind speed and direction over the deck is absolutely crucial and when he comes back onto the deck at night his visual cues are limited to the approach lighting, for direction and glide slope. Once he's in the hover beside the ship, it gives him his fore and aft and lateral cues. In benign conditions, we think that's not too bad. If the deck is pitching, if it's raining, that adds another dimension.

"It's a question of a measured approach to achieving the night capability. We don't expect first-tour pilots to achieve night qualification within their first year or necessarily within their first tour. That's a step too far. Once they are in a position to achieve night qualification, it's a measured work-up to doing it. You don't just 'kick the tyres and light the fires, get airborne in the dark and see how it goes'. You do a launch in day and recover in dusk before it gets dark, and you go through graduated stages of difficulty until you are comfortable. It's a measured process which you have to work on for a long time. It's important that we do that. What it means, of course, is that with the throughput of pilots from the training squadron into the front line, and from the front line to the next stage, whatever that might be, we will have a profile of experience. At the moment that means that we have at least two or three pilots who are on their first tour and are not night-qualified. When you are committed to an operation, and much of it is done at night, there is a certain loading on the guys who are fully capable.

"Not many of our pilots have operated in the Gulf before. Some have. So, of course, the value of the day-only pilots is limited out there, and there is more pressure on the more experienced guys. But after three months or so of flying operations there,

it is surprising how rapidly the new guys come on, using the circumstances of that theatre to best advantage *re* their operational capability. You can bring new guys on quickly. Within a month or so they can be flying as wingmen and the lead of a pair in a day mission. I know the quality of the guys coming out of the training pipeline, and if I can be assured that a new guy coming to us is of that quality, I don't have any particular worries. Fortunately, the quality of the individual is as good as it ever has been and I think that will prevail for some time. So, we can expect the fellow, after a short period, to be ready to meet anything that's asked of him in any operational theatre during the day. When does he start being able to do it at night? That's a big step.

"We deploy to the Gulf early in 2000. I understand that the American carrier *Stennis* will be there when we are. The operational relationship with the American carriers is key. What the American carriers provide is a quantum leap beyond what we can provide from this ship, but the necessary liaison is quite important. In the past we have had one of our fixed-wing squadron guys go over to the American carrier and be a liaison officer, providing the ability to link-in our operation with theirs by having someone face-to-face on station. The other thing is, within the Sea Harrier squadrons we have American exchange officers, and, of course, we have our own guys over in the States on the exchange programme. That is extremely important. In the case of 801 Squadron, the current American exchange guy is a US Marine AV-8 pilot. So we all draw on a wide range of experience, carrier-borne fixed-wing or VSTOL, at-sea experience in a particular mission type. It's all useful and certainly, if you speak the same language, it's jolly helpful. By that, I don't mean the same mother tongue, but the *same language*. If you can talk *fixed-wing aviation* with each other, that's pretty useful. I think the relationship [Royal Navy and US Navy] is quite a close one, but we approach the problem from a slightly different dimension. We like to think that the capability we offer up will be there on the day, and we'll meet our share of the task. I think that has generally been achieved when we have been in the Gulf and the Adriatic. I believe that the Brits and the Americans actually tend to think the same sort of way and there is a keenness to make sure that we and they think about the tasks we have, and the tasks they have (sometimes the same tasks) in the same way, and have the same sort of approach to getting the business done. I think we work well with the Americans.

"When I went through flying training a number of years ago, I suspect there was a degree of latitude that individuals had which I don't think is the case today. I think the standards today are very rigorous and they are applied absolutely. I know the standards and am very happy when a guy comes and joins a squadron here. I know that he is capable. There is less latitude these days and that is perfectly right. We have complicated airplanes and complicated missions to be done, and we need the highest standards for the guys that we are inviting to go off and do these missions. It's tighter these days and I'm glad it is. Not to say that we were getting it wrong twenty years ago, but it was different. The aircraft were older and we didn't have quite so much reliance on technology and switchology which every airplane brings these days. People today have to be extremely capable of making it all work because if it doesn't, the mission is not going to work. It's better today, but we do demand a huge amount from our people and they do come and provide the goods. It's a management challenge to try to keep that motivation and enthusiasm going so that we can sustain."

Airline pilot Paul Ludwig flew AD Skyraiders from the USS *Hornet* (CVA-12) in the 1950s: "Before we left Miramar for a cruise on the *Hornet*, my inexperience in instrument procedures rose up and bit me on the butt. It had been my impression that Miramar [San Diego] always enjoyed CAVU [ceiling and visibility unlimited] weather. San Diego seemed to always have blue skies, but Ops knew that, at the very least, I needed practice using the homer east of Miramar, which hooked up with the GCA [ground-controlled approach]. I was then scheduled for a practice hop, flying wing on a section leader. After receiving a very short briefing, I nodded that I understood the lecture, but really, I hadn't been paying very close attention because San Diego has such nice weather. We took off and I was led to the homer to practise a GCA in absolutely perfect weather. I did the procedure and we landed. I thought I knew the procedure. I didn't. One night, we departed Miramar in CAVU weather and I was flying wing on a guy, just to log some night hours. On our return to base, Miramar was socked in, obscured by low cloud. Over the homer, my flying companion kissed off by blinking his lights, changing to a GCA frequency, and pulling away into the darkness. Where had he gone? Was he still at my altitude? Would he now be on the reverse course to the one I was flying? The thought of a mid-air [collision] crossed my mind. Before changing to the GCA, I fouled up by turning over the homer and putting myself outside the GCA's radar. GCA then complicated my situation by giving me headings and descents even though I wasn't on his radar. He didn't say I wasn't on his radar. I *knew* I wasn't on his radar. My wrong turn caused him to lose me. I then followed radio instructions faithfully and soon broke out in the clear over some foothills. There was a moon, and I was no more than a few hundred feet above hard terrain. As I could now see, I knew I wouldn't fly into the ground as long as the visibility held. I relaxed a little because I had ample fuel. Eventually, GCA told me to climb and head west. When I finally appeared on his radar, he brought me home. I felt stupid. I was stupid. I knew I would have to do better."

operations, a direct dividend of the angled deck revolution, now brought efforts to devise a better, more efficient and powerful means of catapult-launching for carrier aircraft. Previously, planes had been launched by means of hydraulic, pneumatic or cordite-charge catapult systems. With the ever-increasing take-off weight of carrier-borne aircraft, these systems simply were not able to cope, necessitating a new kind of launch device. The new concept took the form of a long cylinder positioned under the flight deck. It was powered by steam from the ship's boilers, which drove the catapult shuttle with enormous acceleration. A bridle arm connected the shuttle to the aircraft through a slot with flexible sealing. A high level of reliability was achieved in early testing aboard HMS *Perseus* and later in the USS *Hancock* (CV-19). The steam catapult has proven reliable and incredibly powerful over the years. The system is, however, quite sensitive and requires considerable care and maintenance to keep the cats operational in the immense demanding flight schedules of the current supercarriers.

The last of the three primary improvements in the design and function of the aircraft carrier is the mirror or optical landing sight. As described in the chapter *LSO*, the system, based on a British idea, has replaced the old role of the landing signal officer and provided a nearly fool-proof means of bringing an aircraft aboard a carrier safely.

KAMIKAZE

I would attack any squadron blockading a port. Nothing could prevent me from dropping out of the clear blue sky on to a battleship with 400 kilos of explosives in the cockpit. Of course it is true that the pilot would be killed, but everything would blow up, and that's what counts.
– Jules Vedrines, French aviation pioneer, pre-1914

The code of the samurai demands that we must always be ready to die, but that does not mean we must commit suicide on the slightest pretext. Our tradition desires that we should live and fight as best we can so as to experience neither regret nor remorse at the moment of death.
– a kamikaze instructor quoted in *L'Epopée Kamikaze* by Bernard Millot

Right: In the US Navy's second worst disaster, 832 crew members of the USS *Franklin* (CV-13) died when the ship was struck by two Japanese bombs during a kamikaze attack on 19 March 1945. The *Franklin* and her crew were the most decorated in US Navy history. Under the command of Captain Leslie E. Gehres, the crew sailed the devastated vessel 12,000 miles to the Brooklyn Navy Yard for lengthy major repairs.

"It is absolutely out of the question for you to return alive. Your mission involves certain death. Your bodies will be dead, but not your spirits. The death of a single one of you will be the birth of a million others. Neglect nothing that may affect your training or your health. You must not leave behind any cause for regret, which would follow you into eternity. And, lastly: do not be in too much of a hurry to die. If you cannot find your target, turn back; next time you may find a more favourable opportunity. Choose a death which brings about a maximum result."
— from *The First Order to the Kamikaze*

IN WHAT MAY HAVE BEEN an incident created by the Japanese military as an excuse for mounting an offensive campaign against the Chinese, an explosive device damaged a vital Japanese rail link in 1931, leading to their incursion into Manchuria, the taking of Peking, and the sacking and burning of Nanking. By 1936 the Japanese Cabinet was dominated by military figures and several senior government officials had been assassinated. A new order had taken control in Japan and the Japanese Imperial Army was on the march through Indo-China. It proceeded to drive the British from Shanghai and the Dutch from the East Indies as the Great East Asia Co-Existence Sphere (the Japanese euphemism for their newly-occupied territories), was being formed.

By early December 1941 Japanese militarists had decided that the time had come to implement a bold, surprise attack on warships of the US Navy at anchor around Ford Island in Pearl Harbor, Hawaii. The raid came in the early hours of 7 December in an attack by carrier-based aircraft of the Imperial Japanese Navy. Their bombs and torpedoes sank the US battleships *Arizona, California, Oklahoma, Utah* and *West Virginia*, and damaged the *Maryland, Nevada, Pennsylvania* and *Tennessee*. Of these, the *California, Nevada* and *West Virginia* were later salvaged. In a great irony,

the Japanese had unwittingly committed an act which would lead to their ultimate destruction as a world military power and to their defeat in World War II. Their attack at Pearl Harbor came at a moment when the principal battleships of the US fleet lay at anchor there, but none of the American aircraft carriers were present. Thus, the Japanese virtually compelled the US Navy to rely heavily on its aircraft carriers throughout the course of the war, the majority of its battleships being unserviceable. Traditionally, the US Navy had led with its big stick, the battleship, but now it would be the aircraft carrier that would form the nucleus of its task force groups.

The Pearl Harbor attack brought the US into the war when President Franklin D. Roosevelt asked the Congress on 8 December 1941 to declare war on the Empire of Japan. In the months to come, Japanese troops moved to capture the Philippines, Singapore and Burma.

In 1942 the US Navy maintained an airfield and a refuelling station on the island of Midway located 1136 miles west of Hawaii. In June of that year Japanese Admiral Isoroku Yamamoto made a plan to eliminate the US carrier threat by drawing the big American ships out for battle at Midway. The Japanese naval force of eight carriers, eleven battleships, eighteen cruisers and sixty-five destroyers significantly outnumbered the American fleet of just three carriers, no battleships, eight cruisers and fifteen destroyers. Still, the Japanese were defeated in the Battle of Midway, 4–6 June (their first naval defeat since 1592), owing largely to the Americans having cracked the Japanese naval codes. The Japanese Navy's failure at Midway marked the turning point of naval power in the Pacific war. Japanese carriers would never again pose quite the same threat to the US fleet, and the US Navy was, from that point, able to go on the offensive. In the Battle of Midway no gunfire was exchanged between the warships. It was an air

battle between planes of the two carrier forces, and the Japanese lost four of their carriers *Akagi*, (which had led the Pearl Harbor attack), *Kaga*, *Hiryu* and *Soryu*, while the Americans lost the carrier *Yorktown*.

In the final months of 1944 Imperial Japan was losing the war, and some of its military leaders began to express the belief that desperate times called for desperate measures. The notion of self-sacrifice for Emperor and country was commonly accepted among the Japanese, and suicide *per se* was not alien, and was honoured for its purity by many who had been raised on tales of heroic Samurai warriors. Thus it was a short step to the concept of suicide as a weapon, i.e. the *kaiten* ("turning the tide") human torpedo. The *kaiten*, 54-feet long, carrying a 3000-pound warhead, had a range of 30 miles at slow speed or 12 miles at a top speed of 40 knots. It was designed around the Japanese Type 93 torpedo, primarily by naval architect Hiroshi Suzukawa. Launching the *kaiten* meant a one-way trip for the crew, who could not get out. Only one US ship was sunk by a *kaiten*, the tanker USS *Mississinewa*, and most *kaitens* proved unstable, killing their "pilots" before reaching their targets.

General Yashida of the Japanese Army Air Force, meanwhile, was advocating suicidal air attacks, and ramming techniques were being secretly included in the pilot training syllabus. Army General Yoshiroko, in command of units in the Solomon Islands, was particularly frustrated by the ineffective anti-tank weapons in his arsenal, and called upon his troops to make the supreme sacrifice by strapping satchels of explosives to their bodies and diving under the tanks of the American enemy. The effort was mostly without result and the General was the subject of severe criticism from Tokyo before being transferred to another post. The concept and use of "human bullets" by the Japanese continued however.

I was walking across the flight deck in the still half-dark dawn to man my plane. I had reached the middle of the deck when I heard the very loud sound of 20mm cannon fire coming from astern the *Randolph*. Almost immediately, huge red tracers began coming up from astern, from below the port edge of the flight deck and parallel to the ship. As I hastily dropped to the deck wishing I could dig a fox hole, a kamikaze Zero arced up alongside the flight deck with his guns still firing, and flashed by less than 50 feet from me. The Zero then nosed down and turned in an attempt to crash into the carrier. Fortunately, he had misjudged his speed and passed just in front of, and under the bow of the ship, crashing and exploding in the water on the starboard side. The carrier got quite a jolt from the bomb explosion and lots of shrapnel on the flight deck, with little damage.
– Hamilton McWhorter, former US Navy fighter pilot, 17 April 1945

Left: From the flight deck of the USS *Enterprise* (CV-6) as the ship sails from Pearl Harbor, Hawaii, early in the evening of 25 September 1945.

On 15 June an invasion force of US Marines landed on the island of Saipan in the Marianas chain, as long-range American B-29 Superfortress bombers were attacking the Japanese mainland. The bombers frequently flew well beyond the range of both anti-aircraft and defending fighters, further frustrating the Japanese high command who were shocked to witness the enemy planes operating, seemingly with impunity, over their sacred homeland.

The First Battle of the Philippine Sea took place on 19 June with Japanese carriers and battleships engaging US naval forces off Saipan. Within this pivotal clash aircraft of both sides met in what has ever since been referred to by US naval aviators as "The Great Marianas Turkey Shoot". When it was over the Japanese had lost 328 carrier-based planes, 50 land-based planes, three more carriers, and the last of their best pilots. The Americans lost just twenty-nine aircraft. After this engagement Japanese naval power was largely neutralized for the remainder of the war.

Vice-Admiral Takejiro Onishi, Imperial Japanese Navy, Chief of the Ministry of Munitions, Arms and Air Control Bureau, was a principal advocate of the kamikaze idea. It was he who originated the name kamikaze, which means "Divine Wind" and is believed to be a reference to the ancient winds that sank the threatening Mongol fleet. Kamikaze pilots were members of special attack units. Their mission was to become human bombs, one with their airplanes, and sacrifice themselves by diving their planes into enemy ships with the goal of sinking them. In an effort to instill high morale among his airmen, the Vice-Admiral introduced some ceremonial aspects to the kamikaze units, including the pre-flight toast of sacred water (later changed to saki), and the wearing of a decorated white headband called a *hachimaki*, a touch of Samurai indicating that the warrior was prepared to fight to the death. The majority also wore a *sennin-bari*, a silk or cloth band stitched with red

threads that was said to have the power of a bullet-proof vest. Most kamikaze pilots carried a personal flag, usually a small square of white cloth with a red *hinomaru* circle in the centre and calligraphy encouraging "a suicide spirit". Kamikaze pilots and their families received privileges, including extra food rations, as well as "very honourable" status. By some, the kamikaze were referred to as "the black-edged cherry blossoms".

Onishi was painfully aware of the shortage of truly skilled Japanese pilots, but he still believed that his aircraft were a potent weapon. "If a pilot facing a ship or plane exhausts all his resources, he still has his plane left as a part of himself. What greater glory than to give his life for emperor and country?"

In his book *Kamikaze—Japan's Suicide Samurai*, Raymond Lamont-Brown states that "the Kamikaze pilots evolved from four main sources of recruitment.

"First came the 'patriotic crusaders' who were all volunteers, usually from *daimyo* or samurai families; they were motivated by nationalistic fervour, military ideals and the concept of chivalry upon which their ancestors had based personal sacrifice to fulfil perceived duties to the state. From this group evolved the ritualization of the kamikaze before suicide flights (i.e., the wearing of samurai symbols, singing patriotic songs, writing poetry glorifying kamikaze action, composing testamentary last letters home, distributing personal effects and so on).

"Next came the 'nation's face savers'. These were recruits who did volunteer, but often for negative reasons, to avoid personal shame in not emulating the deaths of the patriotic crusaders, or to espouse military heroism in order to save the *Kami* land of Japan from humiliating defeat. Like the patriotic crusaders, they too were conformists to the traditions of Japanese society. As the kamikaze Susumu Kitjitsu (1923–45) was to write to his parents: 'I live quite a normal life. Death does not frighten me; my only care is to know if I am going to be able to sink an aircraft carrier by crashing into it.'

Everyone talks about fighting to the last man, but only the Japanese actually do it.
– Field Marshal William Slim, Supreme Allied Commander, SE India in World War II

The psychology behind [the kamikaze attacks] was too alien to us. Americans who fight to live, find it hard to realize that another people will fight to die.
– Admiral William F. Halsey, Commander of US Third Fleet, following the attack on the USS *Intrepid* (CV-11), 25 October 1944

Only the dead have seen the end of war.
– Plato

"By the last few months of the war the third category of recruits emerged: these were the 'young rationalists'. They came me ̃ straight from higher education, went through hurried training and died to sustain the ̃ r effort and to keep Japan free from foreign ta ̃ As Lernard Millot wrote: 'With a few very ̃ ̃ ̃ ̃eptions, they were the most affectionat̃ ̃vell-educated, least troublesome sons who gave their parents the greatest satisfaction.'

"The last group of recruits were also mostly young, the 'appointed daredevils', who emerged right at the end of the war. It may be noted that among their number were do-or-die delinquents, hell-raisers and those of shady moral reputation and social deviation who, through the drastic measure of suicide, were escaping the legal, civic and social consequences of their behaviour."

In pre-war Japan, naval pilots were required to log a minimum of 400 flying hours before they qualified to train for carrier operations. They then had to accrue an additional 400 hours on carriers before being considered combat ready. As the war situation worsened for Japan, its pilot training was reduced to a maximum of 200 hours with virtually no navigation, aerobatics or combat technique included. These poorly-trained student pilots quickly became easy prey for US carrier aviators.

Now the American B-29s were based on Saipan, posing a far greater threat to the Japanese home islands. Japanese High Command, in a progressively more desperate mindset, ordered a dramatic, large-scale, three-pronged battleship and cruiser attack (which did not involve its own carriers) against the US fleet. The plan called for a diversionary force to draw the US carrier-based planes far away from the US fleet. When it failed, the High Command immediately dispatched Vice-Admiral Onishi to the Philippines to take command of the First Air Fleet. US carrier planes, meanwhile, were busy bombing the Japanese airfields there — Clark, Negros and Cebu — causing extensive damage.

Onishi could not actually order his pilots to fly the special suicide attacks — they had to volunteer, and with no expectation of survival, they did so almost unanimously. On 20 October 1944 he addressed twenty-six fighter pilot volunteers who were to comprise the *Shim* (God and Wind) force: "My sons, who can save our country from the desparate situation in which she finds herself? Japan is in grave danger. The salvation of our country is now beyond the power of the Ministers of State, the General Staff, and lowly commanders like myself. It can come only from spirited young men such as you. Thus on behalf of your hundred million countrymen, I ask you this sacrifice, and pray for your success. You are already gods, without earthly desires. But one thing you want to know is that your own crash-dive is not in vain. Regrettably, we will not be able to tell you the results. But I shall watch your efforts to the end and report your deeds to the Throne. You may all rest assured on this point. I ask you to do your best."

Soon after Onishi's address, 201st Air Group *Chusa* Tadashi Nakajima was sent to Cebu in the central Philippines to organize a new kamikaze unit, and told the pilots on his arrival: "I have come here to organize another Special Attack Unit. Others will want to follow in the footsteps of the first pilots charged with this mission. Any non-commissioned officer or enlisted flyer who wishes to volunteer will so signify by writing his name and rate on a piece of paper. Each piece of paper is to be placed in an envelope which will be delivered to me by 2100 hrs today. It is not expected, however, that everyone should volunteer. We know that you are all willing to die in defence of your country. We also realise that some of you, because of your family situation, cannot be expected to offer your life in this way. You should understand also that the number of volunteers required is limited by the small number of planes available. Whether a man volunteers or not will be known only to me. I ask that each man, within the next three hours, come

to a decision based entirely upon his own situation. Special attack operations will be ready to start tomorrow. Because secrecy in this operation is of utmost importance, there must be no discussion about it." All of the pilots volunteered.

The Vice-Admiral began launching his kamikaze missions against US Navy ships on 21 October. That day his pilots failed to locate their targets and returned to their Philippine base, where they had to watch helplessly as many of their precious aircraft were destroyed in a US bombing attack. The outcome was different, though certainly not all good, for Japanese forces the next day. Conventional bombing aircraft of Japan's Second Air Fleet struck at US ships in a massive raid, sinking one carrier and three smaller ships. For the day though, Japan had to endure the loss of three battleships, six cruisers, seven destroyers and more than a hundred aircraft. And by this time, most Imperial Japanese Navy pilots were dead and the role of the kamikaze had to be carried on by Army Air Force pilots, the *Tokko Tai*. On 25 October kamikaze pilots succeeded in sinking a US carrier and damaging several others. In Japan, the Emperor sent his congratulations to the suicide unit on its latest achievement, but disquieting rumours were circulating there that kamikaze claims of destruction and damage of US ships were wildly exaggerated and that perhaps fewer than 10 per cent of such claims were actually valid. A belief seemed to prevail that the pilots of the Divine Wind were reluctant to admit to any sacrifices that had been in vain.

Onishi launched a frenzy of kamikaze activity in November. On the 5th of the month a group of his aircraft was *en route* to strike at a US landing force at Leyte when it encountered a stream of US bombers. All of the Japanese pilots rammed their American adversaries. Then, in an attempt to stave off the US invasion of Luzon, Onishi diverted his pilots from primary attacks on US carriers to hitting transport ships. He also began employing heavy bombers loaded with explosives in his

The question is whether [suicide] is the way out, or the way in.
– from *Journals*
by Ralph Waldo Emerson

Razors pain you, / Rivers are damp, / Acids stain you, / And drugs cause cramp. / Guns aren't lawful; / Nooses give; Gas smells awful; / You might as well live.
– from *Enough Rope*
by Dorothy Parker

We cannot tear out a single page from our life, but we can throw the whole book into the fire.
– from *Mauprat*
by George Sand

Left: In a miscalculated final plunge, a kamikaze narrowly misses the crowded flight deck of the US escort carrier *Sangamon* (AVG-26) in May 1945.

suicide units. His most important achievement of the period came in the third week of November, when his forces flew against the US carriers again, seriously damaging four and causing them to be withdrawn for repairs. The US naval command was then compelled to add many new destroyers as pickets around the carriers, and to double the number of fighters in the carrier air wings for greatly increased combat air patrols. Sailors given shore leave were ordered not to discuss the kamikaze attacks. US saturation bombing of Japanese airfields on Luzon was intensified, leading to the grounding of all locally-operating Japanese aircraft for several days. At this point the weather became a crucial factor, as a typhoon struck the Philippines on 15 December, damaging many ships from the US fleet. Three destroyers were lost and much of the fleet was forced to withdraw for repairs. The fortunes of Onishi's kamikaze units continued to dwindle too. He now had fewer planes than pilots, and decided that, when the enemy forces landed on Luzon, he would order all pilots without planes to fight on as infantry.

On 9 January "Mike One", the US invasion of Luzon, began. The Americans met stiff resistance, including a lot of attention from suicide motor boats as well as the kamikazes. For the crews of the US ships, the only option was to bring maximum concentrated gunfire on the incoming suicide planes. By 13 January the kamikaze campaign had cost the lives of 1208 Japanese pilots. Vice-Admiral Onishi was shifted to Formosa where he quickly organized new kamikaze units. These units were soon to engage a US fleet which had gathered off Formosa as a part of the American effort to take the island of Iwo Jima where it needed to establish a base for Mustang fighters to put them within a range to escort US B-29s on their Japan raids. Japan's urgent goal now was to destroy the US fleet in order to force some sort of honourable peace settlement, and

the kamikaze were key to this objective. In retaliation for the huge B-29 strike on Tokyo of 9 March, which took nearly 100,000 lives and made more than one million people homeless, the Japanese launched Operation *Tan*, a bomber and kamikaze strike on US Navy ships anchored in the harbour at Ulithi. It too failed.

At this point the Japanese position was truly a desperate one. After the fall of Iwo Jima, most in the Japanese military believed that they had but two choices left — surrender or fight to the death utilizing the suicide weapon to the fullest extent. Their anger and frustration at the relentless B-29 fire raids over their homeland reinforced their determination to win the war that they had clearly already lost. They refocused on the kamikaze concept with a grim new dedication. But they were rapidly running out of time, aircraft and fuel, and the forced restrictions on pilot training had resulted in a relatively small corps of airmen who were barely able to fly at all. On 17 March, with the US fleet only 100 miles south of the Japanese mainland, Admiral Yugaki ordered his diminished force of kamikaze and conventional bombers up to strike at the enemy with the greatest intensity. In the attack fifty-two of his aircraft were lost. The US carrier *Franklin* was badly hit and the Americans suffered more than 1000 casualties. Yugaki, however, saw the attack as yet another failure and levelled blame on inferior training practices.

In a display of unusual enmity within the ranks, some kamikaze airmen began referring to their conventional bomber colleagues as "lechers" who seemed to prefer earthly delights over those of the spiritual resting place of their dead heroes. The bomber types, in turn, called the kamikaze madmen.

On 21 March another new development in the *Tokko Tai* programme appeared. Ships of the US fleet were about to be attacked by Japanese bombers carrying manned flying bombs called *Ohkas* (suicide attack aircraft). Before they could release the *Ohkas*, the bombers were spotted by US Navy combat air

Right: The 27,000-ton US carrier *Bunker Hill* (CV-17) burned furiously for five hours after being struck by two Japanese suicide planes on 11 May 1945. Tragically, 660 of her crew lost their lives. The ship was saved by the action of her skipper, George A. Seitz, who ordered an extreme turn which allowed the water that had been used to fight the fires in the hangar deck, to flow into the sea.

patrol aircraft and all of the Japanese planes were shot down.

The island of Okinawa was of major importance to the Americans, as it would provide them with the closest base yet from which to strike with B-29s at the Japanese home islands. As the US fleet, minus some of its carriers that had been significantly damaged in the recent kamikaze attacks and had been withdrawn for major repair work, made its way toward Okinawa, the last major Japanese-held island and the nearest to Japan, an odd thing happened. Japanese reconnaissance reports led them to conclude that the absence of a number of US carriers from the current enemy fleet composition meant that the carriers had been sunk and the fleet was no longer on the offensive. They soon found, however, that this was not the case. The American invasion of Okinawa was moving ahead at full steam. On the morning of 1 April US landing forces arrived on the Okinawan beaches. The Japanese defenders were well established in caves on the island, and lay in wait for the enemy troops who met with little initial resistance during the landings. In the afternoon though, kamikaze attacks began and by 6 April the Japanese had launched what they called "the holy war" against the American enemy.

In 1956 Jean Larteguy's edited version of the George Blond description of a typical kamikaze attack in *Le Survivant du Pacifique* appeared in Larteguy's *The Sun Goes Down*: "On 14 May, at 6.50 a.m., the radar plotter reported an isolated 'blip', bearing 200° at 8,000 feet, range about 20 miles. The rear guns were pointed in that direction, ready to fire as soon as the 'phantom' should appear. At 6.54 it came into sight, flying straight for the carrier. It disappeared for a moment in the clouds; then, after approximately three-and-a-half miles, it emerged again, losing altitude. It was a Zero. The 5-inch guns opened fire. The Japanese aircraft retreated into the clouds. The batteries continued to fire. The crew had been at action stations since

Dear Parents:
Please congratulate me. I have been given a splendid opportunity to die. This is my last day. The destiny of our homeland hinges on the decisive battle in the seas to the south where I shall fall like a blossom from a radiant cherry tree.
– from a last letter home in *The Divine Wind* by Capt. Rikihei Inoguchi and Cdr. Tadashi Nakajima with Roger Pineau

Left: Struck twice within 30 minutes by a Zero and then a Judy, *Bunker Hill* (CV-17) has become an inferno. The ship would survive but would never again see action.

Above: In a kamikaze attack on 25 October 1944, the US escort carrier *St Lo* (CVE-63) was lost when torpedoes stored in the hangar deck exploded, blowing the stern off the ship. Right: A dramatic photograph capturing the instant of a kamikaze strike on the flight deck of a US carrier.

four in the morning. All the aircraft that were not in the air had been de-fuelled and parked below decks.

"The Japanese machine approached from the rear. It was still not to be seen, as it was hidden by the clouds. Guided by radar, the 5-inch guns continued to fire at it, and soon the 40mm machine guns began to fire as well. It was very strange to see all these guns firing relentlessly at an invisible enemy.

"The Japanese aircraft emerged from the clouds and began to dive. His angle of incidence was not more than 30°, his speed approximately 250 knots. There could be no doubt — it was a suicide plane. It was approaching quite slowly and deliberately, and manoeuvring just enough not to be hit too soon.

"The pilot knew his job thoroughly and all those who watched him make his approach felt their mouths go dry. In less than a minute he would have attained his goal; there could be little doubt that this was to crash his machine on the deck [of the carrier *Enterprise*, CV-6].

"All the batteries were firing: the 5-inch guns, the 40mm and the 20mm, even the rifles. The Japanese aircraft dived through a rain of steel. It had been hit in several places and seemed to be trailing a banner of flame and smoke, but it came on, clearly visible, hardly moving, the line of its wings as straight as a sword.

"The deck was deserted; every man, with the exception of the gunners, was lying flat on his face. Flaming and roaring, the fireball passed in front of the 'island' and crashed with a terrible impact just behind the for'ard lift.

"The entire vessel was shaken, some forty yards of the flight deck folded up like a banana-skin: an enormous piece of the lift, at least a third of the platform, was thrown over 300 feet into the air. The explosion killed fourteen men. The last earthly impression they took with them was the picture of the kamikaze trailing his banner of flame and increasing in size with lightning rapidity.

"The mortal remains of the pilot had not disappeared. They had been laid out in a corner of the deck, next to the blackened debris of the machine. The entire crew marched past the corpse of the volunteer of death. The men were less interested in his finely modelled features, his wide-open eyes which were now glazed over, than in the buttons on his tunic, which were to become wonderful souvenirs of the war for a few privileged officers of high rank. These buttons, now black, were stamped in relief with the insignia of the kamikaze corps: a cherry blossom with three petals."

The men of the US Navy ships off Okinawa were surprised and amazed by the numbers and ferocity of kamikaze and escort aircraft subjecting them to this new and greatly intensified attack. These American sailors who were fighting to live were up against Japanese airmen determined to die . . . a shocking realization that was repeatedly brought home to them during the incessant attacks. In addition to the efforts of the kamikaze, Japanese warships, including *Yamato*, the world's largest battleship, sailed to Okinawa in an attempt to destroy the US transport ships. They then intended to deliberately run the *Yamato* aground and use her as a coastal fortress. The Japanese plan was foiled, however, when US Navy PBY Catalina flying boats sighted the Japanese warships, and US carriers launched air strikes sinking the *Yamato*, the *Yahagi* and four smaller ships. By 19 June the fight for Okinawa had reached a critical state for the Japanese defenders, and their commander ordered all of his troops to "go out and die". When Okinawa had fallen to the Americans after 82 days of fighting, they had suffered 12,500 casualties; the Japanese more than 100,000. A total of thirty US ships had been sunk. In Japan the people were being told that every citizen was now considered a *Toko Gunjin,* or special attack soldier, for the defence of the homeland.

In the Marianas, the B-29s of General Curtis LeMay's 20th Air Force were continuing their campaign of fire raids against Tokyo and the sixty or so prinicipal cities of Japan. It had begun in

In his official code of ethics of January 1941, then Army Minister of Japan Hideki Tojo said: "Do not think of death as you use up every ounce of your strength to fulfil your duties. Make it your joy to use every last bit of your physical and spiritual strength in what you do. Do not fear to die for the cause of ever-lasting justice. Do not stay alive in dishonour. Do not die in such a way as to leave a bad name behind you."

Right: Made famous as the host ship in the Doolittle Tokyo Raid of 18 April 1942, the USS *Hornet* (CV-8) was badly damaged on 26 October 1942 when it came under attack by two apparent suicide planes. Severely damaged in the attack, *Hornet* was later sunk by the Japanese destroyers *Akigumo* and *Makikumo*. Below and right: Suicide attacks on the US carrier *Enterprise*, 14 May 1945.

January when LeMay took command of the 20th and reorganized it for the task of efficiently bombing the Japanese into submission and, hopefully, bringing the Pacific war to an end.

The American raids were still going on in early August when the US President, Harry S. Truman, ordered the use of the first atomic bomb on a target city in Japan. This first use of a nuclear weapon in war took place on 6 August when a B-29 called *Enola Gay* (named after the mother of the pilot and airplane commander, Colonel Paul W. Tibbets, Jr.) released the bomb called *Little Boy* over the city of Hiroshima. *Little Boy*, together with the bomb called *Fat Man* which was dropped on Nagasaki three days later, led to Japan's capitulation, but not before Russia declared war on Japan on 8 August and launched its own offensive in Manchuria, completely overwhelming the Japanese there.

In what would be the final mission of the kamikaze airmen, Vice-Admiral Matome Ugaki launched what remained of his force against the US carriers on 15 August. On the way to the target he sent this radio message to his headquarters: "I alone am to blame for our failure to defend the homeland and destroy the arrogant enemy. The valiant efforts of all officers and men of my command during the past six months have been greatly appreciated. I am going to make an attack at Okinawa where my men have fallen like cherry blossoms. There I will crash into and destroy the conceited enemy in the true spirit of Bushido, with firm conviction and faith in the eternity of Imperial Japan. I trust that the members of all units under my command will understand my motives and will overcome all hardships of the future and strive for the reconstruction of our great homeland that it may survive forever. *Tenno Hai Ka. Banzai!*" That evening Vice-Admiral Onishi took his own life.

Wilbur "Nick" Nechochea was assigned to V-1

Division as a fire marshall aboard the USS *Enterprise* (CV-6) in World War II. "The *Enterprise* had been hit just before and just after I was on board, but never did sustain serious combat damage during my tenure. That, however, doesn't mean that she and I didn't have some close calls together. On several occasions I observed torpedoes cross the bow of the 'Big E' and we were one of the first Task Forces to be attacked by kamikazes. One of the things I'll always remember about the kamikazes was that with a regular bomber the AA gunners would switch to a new target after they saw the bomb release. With the kamikazes that never happened. The gunners just kept firing at them all the way down. Several times I would do my best to crawl into the finger fittings on deck when we were under attack."

On Kudan Hill in the heart of Tokyo, near the Imperial Palace, stands the *Yasukuni-jinja*, or "Shrine for Establishing Peace in the Empire". It is dedicated to Japan's war dead and is a controversial memorial because it contains personal effects of executed war criminals, including Hideki Tojo. "Even today," according to Raymond Lamont-Brown, "any government minister who makes an official visit to the shrine would be technically liable to be stripped of his office." Displays in the *Yasukuni* include relics of the kamikaze pilots of the Great East Asian War, as the Japanese refer to World War II. Japanese war veterans groups, as well as representatives of the Bereaved Families Association, regularly visit the shrine and petition the public to sign requests for the *Yasukuni* to be reinstated as the official Japanese war memorial. Lamont-Brown states: "As time passes, according to some sections of the Japanese press, the spirits of the dead kamikaze 'cry out' for honourable, official recognition through the members of the 'Thunder Gods Association' who meet annually at the *Yasukuni-jinja* on 21 March (the day on which the first *Ohka* suicide attack was made)."

Is it sin / To rush into the secret house of death / Ere death dare come to us?
— from *Antony and Cleopatra* by William Shakespeare

In the last days before their final attacks, the kamikaze pilots were mostly calmed by the Bushido philosophy. They were able to relax in a seeming detachment, spending their waiting time listening to gramophone records, playing cards, reading, and writing their last letters home. They gave their belongings to comrades and friends, and they all carried 3 *sen* in copper coins, their fare to cross the Buddhist equivalent of the River Styx.

THE AVERAGE AGE of those responsible for carrying out the many vital tasks which keep things running smoothly on the flight deck of a US supercarrier, is nineteen.

On a US Navy carrier, the men and women who do the jobs that make flight operations possible are easily identifiable by the colours of the jerseys they wear. BLUE jerseys are worn by airplane handlers, tractor drivers, aircraft elevator operators and messengers/phone talkers. Air Wing plane captains and Air Wing line leading petty officers wear BROWN. Those wearing GREEN are the catapult and arresting gear crews, Air Wing maintenance personnel, Air Wing quality assurance (QA) personnel, cargo handling personnel, ground support equipment trouble-shooters, hook runners, photographers, and helicopter landing signal enlisted personnel. The aviation fuel personnel wear PURPLE. Ordnance men wear RED, as do crash and salvage crews, and explosive ordnance disposal (EOD) personnel. WHITE jerseys are worn by squadron plane inspectors, landing signal officers (LSO), air transfer officers, liquid oxygen crews, safety observers, and medical personnel. YELLOW jerseys signify aircraft handling officers, catapult and arresting gear officers, and plane directors.

In the aircraft carriers of the Royal Navy, flight deck personnel wear a coloured vest called a surcoat which, as with the American jerseys, identifies personnel by function. YELLOW means Flight Deck Officer, Chief of the Flight Deck, and aircraft directors. BLUE surcoats are for naval airmen/flight deck, and photographers. Aircraft and engine full supervisory ratings wear BROWN. Air electrical full supervisory ratings wear GREEN. GREEN WITH A BLUE STRIPE indicates air radio full supervisory ratings. Crash and salvage parties wear RED. RED WITH A BLACK STRIPE is worn by weapon supply/all ratings. Flight deck assault guides wear RED WITH A WHITE STRIPE. Medical attendants wear WHITE WITH A RED CROSS.

KEEPING 'EM UP

It's always a good idea to keep the pointy end going forward as much as possible.

Left: Flight deck personnel on a US supercarrier manhandle an F/A-18 into a parking spot.

Deck supervisors, duty aircrew, watch chiefs, and Air Engineering Officers wear WHITE. WHITE WITH A BLACK STRIPE is worn by flight deck engineers. Aircraft engine mechanics (all trades) wear GREY.

When a US aircraft carrier is to conduct flight operations, preparations normally begin the day before. An Air Plan outlining the scheduled activity is prepared and distributed the night before. It includes all the required information for those concerned, including launch and recovery times, and information on the mission itself: the number of sorties to be flown, fuel and ordnance load requirements, and the tactical communication frequencies to be used. Flight quarters are announced and manned. No crew members who are not directly involved in the flight operations are allowed to be on the flight deck or in the deck edge catwalks.

The pilots and aircrew are briefed on the specifics of their mission and on the sequence of events. They will go to their aircraft 30 to 45 minutes before they are scheduled to launch, and will do a thorough preflight inspection of the aircraft prior to the order to start engines.

Meanwhile, the various flight deck personnel are preparing their equipment and personal gear for the upcoming ops. Before any aircraft engines are started, a ritual FOD (Foreign Object Damage) walk-down is conducted, in which all off-duty personnel, mainly flight deck and Air Wing, are requested to participate. Often the walk-down is sponsored by one of the Air Wing squadrons, with music provided for motivation. With the range of aircraft minor maintenance and repair activity occuring on the flight deck, it is inevitable that items such as small tools, rivets and bits of safety wire hit the deck and are not noticed at the time. When aircraft engines are running, these objects can be blown about and cause significant injury to people and engines. When flight deck ops are not underway, ship's personnel are frequently allowed to exercise there, usually by running laps around the deck. They, too, can be a source for foreign objects finding their way to the flight deck. The potentially deadly objects are often hidden in the recessed "pad-eye" tie-down points spotted all over the flight deck. Personnel manning air hoses precede the main walk-down force to blow any collected debris or water from these pad-eyes. The deck-wide line of FOD walkers proceeds slowly down the whole length of the flight deck, picking up all objects that may pose even the slightest threat to man and machine. They are followed by scrubber vacuums which suck up anything that may have been overlooked by the walkers. FOD is a deadly menace to both men and machines and the walk-down procedure is taken seriously. Only after it has been completed is the order to start aircraft engines given.

The first craft to start up and launch are the plane guard helicopters who leave the deck to orbit in a D-shaped flight path, designed to let them quickly rescue an airman or crewman should that be necessary.

Just prior to the launch of the mission aircraft, the supercarrier is turned to a heading that will allow for sufficient wind, usually about 30 knots over the flight deck, to assist the planes in getting airborne. The yellow-shirted plane directors begin to guide the first aircraft to be launched to precise spots on the two forward steam catapults. When spotted there, large blast deflectors rise from the deck just behind the planes, to protect deck personnel aft of the catapults. Hook-up green shirts crouch at the nose wheel of each of the aircraft on cats one and two and attach the nose gear to the catapult shuttle with nose-tow and hold-back bars. A green shirt moves in to the right of one plane's canopy and holds up a black box with illuminated numerals which flash the predicted weight of the aircraft. The pilot must concur that the figure is correct. That done, the cat launch personnel calibrate the power of the cat to the requirement of the plane about to launch. Low clouds of steam billow down the length of the cats as a yellow shirt signals the pilot, who releases the brakes and applies full power. The cat officer signals with a rotating hand, two fingers extended, as the pilot does a quick final check that the aircraft and controls are functioning correctly. The pilot then salutes to indicate that he is ready to launch, and braces himself. If he is flying an F-18 Hornet, procedure requires him to place his right (stick) hand on the canopy frame grab handle and keep it there for the duration of the cat shot, in order not to disturb the computer-set trim during the shot.

At this point the catapult officer checks the final readiness of the cat and receives confirmation from other deck personnel that the aircraft is ready for flight. He then signals the shooter in the enclosed launch station bubble (by touching the deck), to press the cat firing button.

Launch . . . and the four-g force of the steam cat hurls the airplane from the flight deck. The plane achieves 150 knots air speed from a standing start in two seconds, sending the flesh and facial muscles of the pilot racing towards the back of his skull.

The first two aircraft of the mission have departed and the catapult crews rush to position and attach the next planes in the queue for launch. These crews can ready and launch an aircraft every 30 seconds if necessary, and are frequently required to set up and launch more than 100 times in a day. For the flight deck operation to go smoothly, an endless regime of planning, discipline, expert engineering, skilled maintenance, training, motivation, and extraordinary attention to detail is employed. And safety is the prime concern of all.

Chris Hurst, a Leading Aircraftman/Aircraft Handler with the Air Department aboard HMS *Illustrious*,

Below: Deck crew aboard a Royal Navy carrier early in the 1940s.

Above: A catapult crewman at the launch of an F-14 Tomcat from a US supercarrier. Above centre: An F/A-18 Hornet aircraft director at work. Above right: A FOD walkdown on a US Navy supercarrier. Right: This catapult crew member displays the weight of a cat-mounted jet that is ready for launch. The pilot or another member of the flight crew must concur with the weight figure and signal that OK to the cat crewman. The cat crew can then precisely calibrate the power of the steam catapult that will launch the jet at roughly 160 miles per hour. Far right: Air Wing personnel in a pre-launch mood.

describes the Sea Harrier launch sequence: "Ten minutes prior to launch, we will get a verbal communication from the flight control position to start the aircraft. Permission will then be given to the aircraft mechanics to liaise with the pilots to start the engines and go through their various acceleration checks, making sure the wing flaps, etc, are at the right angles for launch, depending on the aircraft weights and weapon loads. The mechanics will also take off the outrigger ground locks and the lashings that are not required. Then they will be ready on deck.

"The aircraft directors will face towards the flying control position, watching for an amber light which means that the ship is on a designated flying course and we have permission to taxi the aircraft onto the runway. The pilot will be told to 'unbrakes'. The two remaining nose lashings and the chocks will be removed. Then a Leading Aircraftman will guide the aircraft out of the range and pass it on to the Petty Officer of the Deck who is standing on the runway at the designated launch distance. He will then marshal it onto that launch distance, stop it on the brakes and pass the control of the aircraft launch to either the Captain of the flight deck, or the Flight Deck Officer, whichever is

134

After reinstallation, every jet engine was thoroughly checked out, and the more "Walter Mitty-like" mechanics, including me, eagerly looked forward to this task. One did feel slightly elated, clambering into the cockpit, starting the engine and testing it. While stationed at land bases, we sometimes needed to taxi the aircraft short distances, and one couldn't help wondering how it would feel to actually take off.
– Bill Hannan, former US Navy jet engine mechanic

on watch at the time.

"The duty squadron Air Engineering Officer will then look round the aircraft and make sure that everything is safe, all the relevant pins have been removed, etc. When he is satisfied, he will give a thumbs-up to the Flight Deck Officer, who will wait for a steady green light from Flyco which means he has the Captain's permission to launch fixed-wing aircraft.

"Then, when he has checked up and down the runway that all is clear, he will raise a small green flag. The pilot will turn on a white nose wheel light and roar away up the deck and off the ski-ramp. As that aircraft is launched, the next aircraft will be drawn out of the range and marshalled on in sequence until all the aircraft have gone. All the time this is going on, there is a spare aircraft handling team ready in the 'graveyard' at the front end of the deck, should anything go wrong. We call it the graveyard because it is for dead aircraft. They have a tractor ready to attach to the aircraft. Certain minor unserviceabilities could mean an aircraft not launching, but having to taxi all the way up to the graveyard to get it out of the way, to

clear the deck and make everything ready for the next aircraft to launch. There is nothing worse than an aircraft having a minor radio problem, sitting there with all its intake blanks missing. It would be dangerous. Foreign object damage could occur with an aircraft zooming up the deck, so we prefer to get him straight to the graveyard and out of the way."

"The *Enterprise* flight deck was 109 feet abeam (including the island) and approximately 800 feet long. Usually less than half of that length was available for Scouting Six, because the TBDs of Torpedo Six and SBDs of Bombing Six were always in the pack behind us.

"We frequently watched a bomb-laden SBD drop out of sight as it took off and passed the bow of the ship. It then reappeared, picking up speed, getting a boost from the 65-foot deck height and the 'ground effect' between wings and water.

"One day I was up on deck watching as a young pilot really almost touched the waves ahead of the ship. He later confided that he had taken off with

Right: Aviation Machinist Mate Bill Hannan of VF-721 with an F9F Panther aboard the USS *Kearsarge* in 1952.

his controls fully locked. Somehow he managed to remove the unlocking pin under the control stick, barely avoiding a crash into the sea. He must have been a contortionist."
– Jack "Dusty" Kleiss, former US Navy pilot

Airline pilot David Smith is a veteran of more than 1000 hours of flying the Grumman F-14 Tomcat between 1982 and 1991. In that time he made 342 carrier landings in the course of two Mediterranean cruises aboard the USS *John F. Kennedy* (CVA-67). He was then assigned as an instructor, training fleet F-14 and F-18 pilots in adversary flight tactics at Key West, Florida. All of the F/A-18 Hornet squadrons that took part in the Gulf War Operation *Desert Storm* attended the Key West course: "We F-14 pilots flew whenever the ship had air operations. In a normal schedule we flew during the day and every other night. Some of us flew every night to make up for some of the weak night pilots and the senior squadron officers who just didn't want to fly at night. Flying at night was not fun. Flying in the day was, and everyone wanted to do that.

"Most flight ops lasted either one hour and thirty minutes, or one hour and forty-five minutes. In peacetime, the briefs started between one-and-a-half and two hours before launch. The topics included such admin items as weather, aircraft, crews, join-up (rendezvous), lost communications, gas (in-flight refuelling amounts, altitudes and locations), and the mission briefing itself, which took longer and varied from ship support, to air-to-air basic fighter manoeuvring, to air intercept, to strafing, to section or division tactics. If operating near land, we might be supporting an overland operation or working with another country. The brief was where you transitioned from just being aboard the ship, to being reminded of why you were there. The actual brief was almost a formality because you had heard it all many times before, but it always served to make me focus on what I was about to do.

"If the brief ran long, you would walk immediately to Maintenance Control to check out the logbook on your airplane. You then signed for the airplane and went to the Paraloft, where flight equipment was stored. On the *Kennedy* it was a small room, no bigger than an average bedroom. With all of the hanging equipment, it could barely accommodate four persons. This was where you became an aviator. You went about the task of suiting up and preparing to go topside to do something that no one else on the planet was going to do at that moment.

"On the flight deck it was usually very quiet and rather peaceful at the beginning of a new ops cycle. Normally it was a very dangerous place.

"The first task was to find your plane. There was never a clue as to where it would be spotted. You walked around the flight deck until you found the aircraft number you were looking for. People have been known to man the wrong airplane. Finding your airplane at night can be difficult and frustrating. When you found it, the plane captain (who might be just eighteen years old) would exchange a few words with you about the only thing you had in common: that airplane. I knew all my plane captains and all the maintenance personnel very well.

"Pre-flighting an F-14 on board a carrier can be the most dangerous part of the mission. Some of them were parked with their tails hanging out over the edge of the ship. There could be 20 to 30 knots of wind over the deck, and trying to do a thorough pre-flight might put you in the safety nets, if not actually overboard. Walking around the plane you encountered chains that were trying to trip you, missile fins aiming to leave an impression on you, and many other little gotchas out to ruin your day.

"Climbing in and sitting in the seat of the Tomcat was a relief, a place of comfort. You knew your way around the cockpit with your eyes closed, and you felt safe. Because the F-14 burns gas fast, we always

One afternoon while Ensign Willie P. West and I were walking and talking to each other on the flight deck, neither of us heard the centerline elevator warning signal — if it was sounded. Suddenly, Willie took a step into space, and I was right on the edge of the gaping hole.

I expected to find him in a crumpled heap at the end of a 30-foot drop. Instead, he walked away unhurt. He said that the elevator was moving downward almost as fast as he was falling, and that jumping on it was like landing on a feather bed.
– Jack "Dusty" Kleiss, former US Navy pilot

launched right after the E-2 Hawkeye (our eyes and ears) and the A-6 Intruder/tanker (our gas). A couple of F-14s would be spotted either on the catapults or just behind them. When the order was given, engines were started and we went through the various after-start checks.

"We had afterburners, but did not always need them for the cat shot. On deployment, when you are going to launch missiles and have a full load of fuel, you would always be using afterburners in the launch. You followed the plane director's instructions and he would steer you into alignment just behind the catapult. He would then point to another plane director who was straddling the cat track at a point about 30 feet ahead of your plane. He would slowly lead you forward until you were just behind the cat shuttle. Another director would guide you very precisely onto the shuttle and you would then be turned over to the CAT officer. When satisfied that your hookup was correct, he would have the tension taken on your plane. Under tension, you would be required to go to full power. You did not come out of full power under any circumstances unless the CAT officer stepped in front of your plane and gave you the 'throttle back' signal. This was a trust that had been violated in the past, costing lives. Once under tension and in full power, you were going flying.

"During the day you can see all of the actors on the stage. At night you see nothing but the yellow wands. Two totally different worlds. The difference is almost indescribable.

"The plane is under tension and at full power and, when the CAT officer is content that all of the check-list items have been checked, he will give the signal to launch by touching the deck. Now the fun begins.

"Our initial indication of the launch is when the launch bar releases. F-14 pilots have long been accused of trying to be 'cool' by holding the head forward during a launch, instead of keeping it back against the headrest as most aviators do. Actually, when that launch bar releases, the airplane abruptly squats; it's almost like hitting a pothole in the road. Your head is forced down and forward for a split second. This is quickly followed by an immediate and, hopefully, incredible acceleration forward. The neophyte who puts his or her head back against the headrest at the start of this sequence, will have it jolted forward, and then back, with amazing sharpness that will literally make him or her see stars.

"The intensity of the acceleration can vary, depending on the initial gross weight of the plane and the natural wind over the deck. You need a given end/air speed to go flying, and the heavier your plane is, the greater that air speed has to be.

I was wired to the port catapult of the *San Jacinto*, when the boing boing boing went off followed by 'This is no drill.' I saw a Japanese Jill aircraft coming across the front of our ship, starboard to port. He was low, about 100 feet, and every ship in the fleet was shooting at him. I started yelling at the catapult officer to launch my plane. It seemed to take forever. I charged my guns while waiting. The Japanese plane had bore-sighted another carrier to our port side. Finally, the cat fired. The Jill, with bomb or torpedo in plain view, was crossing my launch path. I squeezed the trigger. Nothing. The gear was still down. The guns are not supposed to fire with the gear down. I hit the gear-up lever. The Jill was closing at full deflection. I was squeezing the trigger. At last, all six .50s were firing just as the Jill crossed in front of me. It exploded as it flew into my line of fire. I had been in the air less than thirty seconds. My aircraft was hit several times by shrapnel, but I continued my combat air patrol and landed just under four hours later.
– James B. Cain, former US Navy fighter pilot, 19 March 1945

Less wind over the deck requires a harder, more powerful cat shot. At a maximum gross weight of 72,000 pounds, an F-14A, with little or no wind over the deck, requires a cat shot that will take the breath out of you. The acceleration is so rapid, it hurts. Such cat shots are not only hard on the pilot and RIO; they are hard on the plane as well. Generators, inertial navigation system alignments, radars . . . all are in jeopardy in such launches. Stories of instruments coming out and striking air crew in the face and chest are not uncommon.

"On an extremely windy day, when the ship is barely moving, you may become concerned that your launch will not be hard enough to send you flying. That concern is directed towards the CAT officer, who may be about to shoot you into water breaking over the deck. He will try to time the launch with the ship's pitches, but a cat shot into a severe down cycle will give you a windscreen full of ocean that you won't soon forget. Still, when things work, and they almost always do, the cat shot is the most enjoyable part of the cycle. Getting back on board is something else."

"Cat shots never bothered me. You are as much passenger as pilot. Even if you shut down the engines and set the parking brake, you were going off the cat. Just 5 knots slower. There is no way to practise cat shots. The first one is the first one. It was memorable and very exhilarating. My subsequent shots were in unison with hooting and hollering, maybe another reason why instructors don't ride along. Once you've finished your required number of traps, you sit on the deck, refuelling and anxiously waiting for the radio call from the LSO and the magic words: 'You're a qual.' Back on land, we were all very excited and animated. I had been in the Navy for only a year, and flying for only eight months. That night I slept like a baby. The next step was advanced jet training."
– Frank Furbish, former US Navy fighter pilot

"One day when I was flying wing out over the ocean west of Miramar, my engine began making grinding sounds. I asked my section leader to take us home. When we arrived, I wrote up the bad engine. The crew chief refused to believe my write-up because I was the new kid, the most junior officer. I tried to convince him about what I had experienced, and left the Ops shack to return to the squadron area. Soon after I got there, the chief phoned me to chew me out, saying I was wrong, that it had only been carburettor icing. Obviously he didn't want to go to work checking the oil filter on the opinion of a snot-nosed ensign. I was just a kid and he an old chief. My mistake was in not contesting his judgement or challenging his lack of respect. He released the plane for ops without checking the oil filter. The next day a squadron buddy of mine flew that same Skyraider and suffered engine failure on take-off, but managed to get it back on the runway. The filter had metal particles in it. The chief did not apologize for his stupidity."
– Paul Ludwig, former US Navy attack pilot

"We had just left Majuro anchorage. We were on our way to Hollandia. Due to strike there on 21 April. We were getting close to Hollandia, in an area where strict security required a darkened ship. AOM2/c Petty had left Fighter Armory with AOM1/c F.S. Rice, heading toward the port bow. It was one of those nights when you literally could not see your hand in front of your face. Both [men] were cautious and felt somewhat safe because they thought the 'safety chain' was up across the bow. Neither realized that we had spotted two F4U night fighters at 'ready' on the catapults, [with the] pilots in them prepared for instant take-off. The safety chain was down.

"Petty was a step or two in front of Rice when suddenly Rice heard a muffled grunt. It didn't take him long to realize that Petty had walked off the bow. Rice immediately ran aft to the LSO's platform where he grabbed some float lights, and without thinking of the consequences, threw some of them overboard. What a sight to suddenly see lights behind the ship in that total darkness. Alert men on watch were quick to report. GQ was sounded and this particular GQ caught everyone at a time when 'tense' didn't begin to describe how jumpy we were.

"The Officer of the Deck sounded 'man overboard' and every division met for immediate muster and roll call. 'Tin Cans' [destroyers] guarding the rear of the Task Force were given the OK to search for Petty. They directed their efforts to the float lights, but there was only one thing wrong with that. The length of flight deck that Rice had to travel to get to the LSO platform would take a fast runner forty to sixty seconds. The ship was moving at a fair rate of speed. In the time he took to run that distance, the closest light to Petty was judged to be a half-mile away. Seas were also running choppy at 4 feet. The odds of finding Petty were in the needle-in-the-haystack category. We were thinking we would

never see Petty again.

"However, there are survivors, some of which are extremely lucky, and some personally self-sufficient. Petty was both. The whale boats which the Tin Cans sent to look for him did an ever-expanding circle search out from the float lights. They had just decided to halt the search when they heard Petty whistle. He could not see them, but he heard their voices, so he managed to whistle with his fingers to his teeth. That is what saved him. He also helped himself by tying knots in his dungarees and filling them full of air as a float.

"We were back at Pearl in January 1944. Petty had gone on liberty and come back with tattoos of a rooster on one foot and a pig on the other. Old seaman's lore has it that a sailor who does that will never drown. When we next returned to Pearl in July, we had sailors by the dozen getting tattooed with pigs and roosters on their insteps."
— M.S. Cochran, formerly assigned to the USS *Enterprise* (CV-6)

Below: "Minsi II", the F6F-5 Hellcat fighter of the *Essex* Air Group boss, Commander David McCampbell, undergoing maintenance on the flight deck of *Essex* on 30 July 1944, off Saipan.

AIR STRIKE KOREA

FROM JUNE 1950 UNTIL JUNE 1953, a war called "a police action" raged on the Korean peninsula, during which attack missions flown by US and British naval aviators took a huge toll on the military assets and the infrastructure of their North Korean enemy. The pilots destroyed 83 enemy aircraft, 313 bridges, 12,789 military buildings, 262 junks and river craft, 220 locomotives, 1421 rail cars, 163 tanks and nearly 3000 support vehicles. The Korean conflict was one of only two events in which United Nations forces have gone to war against an aggressor nation. It was only the absence of the Soviet Union from the UN Security Council (which it was boycotting when war broke out in Korea) that enabled the UN to approve military action there under its auspices, as the Russians would certainly have exercised their veto had they been present.

John Franklin Bolt flew US Marine Corps F4U Corsairs in the Pacific during World War II. He came home with six Japanese Zeke fighters to his credit, and a few years later was to become the only Navy Department jet ace of the Korean war, having disposed of six MiG-15s in air action there.

Jack Bolt wanted a piece of the Korean action. "I chose an Air Force exchange tour of duty because at that point the only thing standing up to the MiGs were the F-86s. I knew that none were coming to the Marine Corps, and I was anxious to get back to the air-to-air fight. The only possibility of doing that was by getting in an F-86 squadron. The MiGs were beating the hell out of everything else, and the F-86s were our sole air superiority plane in Korea from early on. So I managed to get a year's tour with the Air Force, and toward the end of it I managed to get into an F-86 squadron of the Oregon Air National Guard.

"I would grind out the hours in that thing, standing air defence alerts. Those were the days when the threat of nuclear war with Russia hung heavily over our heads. We really thought we were going to get into it. It was late 1951 or early 1952

Left: An F9F-6 Panther of VF-24 on the US carrier *Yorktown* (CV-10) ready to launch on a Korean War air strike in 1953. Above: US Marine Corps pilot Jack Bolt on the wing of his Korean War Sabre.

and the squadron was part of the Northwest Air Defense Command. Our guys were on standby and would be down in the ready room in flight gear, not sitting in the planes the way they did later. They would be playing bridge and so forth, but I would be up flying, getting hours in that F-86. When my Guard tour was over, I was able to get out to Korea in about May of '52 and flew ninety-four missions in F9F-4 Panthers, interdictions, air-to-ground, and close-air support. We were down at K-3, an airfield near Po Hang Do, with VMF-115.

"We were using napalm to attack rice-straw thatched-roof targets. They were all supposed to contain enemy troops but I'm sure most of them were probably innocent civilians. We were attacking the villages with napalm. Then, of course, the proper military targets used thatching for waterproofing too, a supply depot for food or materiel, fuel or ammo. We used lots of napalm against all of those targets.

"My tour in Panthers came to an end and I took some R and R. I looked up an Air Force squadron

Left: The US carrier *Midway* (CVA-41) operating on 18 May 1951. Below: US Navy fighter pilots relaxing between missions in the Korean War.

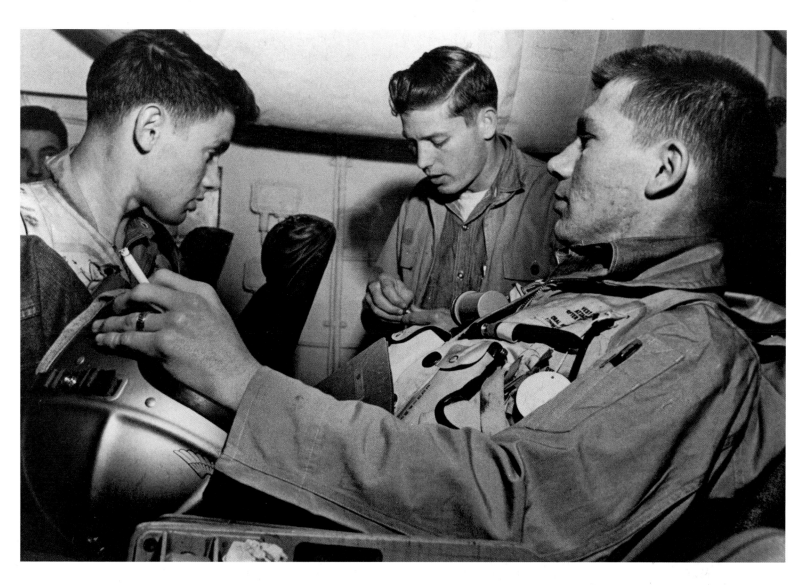

145

Above: A horrifying photo sequence in which a McDonnell F2H Banshee fighter, in a landing approach to the flight deck of the USS *Oriskany* (CV-34), strikes the ramp, explodes and burns. The Banshee entered combat in the Korean War in August 1951 with VF-172 flying from the USS *Essex* (CV-9), escorting USAF B-29 bombing raids. The Banshees attacked rail lines and bridges as well as operating in an escort role with the bombers.

commander named George Ruddell. I had met him at El Toro in 1947 when he flew an F-80 over to lecture our squadron about it and gave us a little demonstration of what that airplane could do. I found George at K-13 where he was commanding the 39th Fighter Interceptor Squadron. I told him about my 100 hours in the F-86F, the same type of Sabre his squadron was flying. I had the experience he needed, and he was friendly towards me and let me take a few fam flights with some of his boys.

"On a second R and R trip around Christmas of '52, it happened that Joe McConnell, who was to become the top-scoring US fighter ace of the Korean war with sixteen kills, had just been grounded from operations, and Ruddell very generously sent him over to teach me some tactics. I flew a few fam hops with McConnell and he was good. We became friends and he taught me a lot about his tactics in the F-86. He was very deserving of the fame that he had earned as the leading ace of that war. Tragically, he was killed soon after the war on a test flight at Edwards Air Force Base in California.

"After that second R and R, I put in for another Air Force exchange tour. The Group Personnel Officer

said, 'Bolt, I know you've been trying to worm your way into this, going up there on R and R. I'll tell you, you've had a year with the Air Force and you ain't goin' up there. You think you are. I'm telling you now, it ain't gonna happen.' They felt that I'd had more than my share of gravy assignments, so I got in touch with Ruddell and he got the general up there to send a wire down to the Marine Corps general. They only had two F-86 groups, the 4th and the 51st, and they had two Marines in each. One of them, a guy named Roy Reed, was leaving shortly. This was the opening I needed. The Air Force general's wire read: 'We're willing to have your pilots, but they come up here having never flown the plane, and they present a training burden on our people. But now we have a rare instance of having a pilot who's shown enough initiative to come up here and get checked out, and he's ready to go. Would you mind appointing John Bolt?'

"There was nothing else the Group could do. I was put in Ruddell's squadron and I was flying on McConnell's wing for my first half dozen flights. I was in Dog Flight. Ruddell was a very tough guy, but he was as nice as he could be. He had four or five kills, but the MiGs had stopped coming south of the Yalu River and we weren't allowed to go

north of it. The Chinese were yelling about the 'pirates' that were coming over there, but that's where the action was.

"When McConnell left, I took over the command of Dog Flight, a quarter of the squadron with about twelve pilots. We lived in one big Quonset hut.

"Ruddell wasn't getting any MiGs because they weren't coming south of the river. He'd been threatening everybody that he'd kill 'em, cut their heads off if they went north of the river after MiGs. But one night he weakened. He'd had a few drinks and he called me into this little cubbyhole where he had his quarters. During the discussion tears came to his eyes — running down his cheeks — as he was saying how he wanted to be a good Air Force officer, and he loved the Air Force, and if they told him to do something, he'd do it, and if they told him not to do something, he'd not do it. But getting those MiGs meant more to him than his career and life itself. And, since he had been beating up on his own flight about not going across the river, he'd be embarrassed to ask any of them to go across it with him. He didn't know whether they would want to anyway — two or three members of Dog Flight didn't like to do it. (They would have been in big trouble if they'd

been identified as going up there. I don't know if the ones they picked up later on, who were shot down north of the river, were ever disciplined when the war was over. But at the time, the threat was believed and hanging very heavily over us.) Ruddell said. 'Would you give me some of your flight? I want to go across the river. I've gotta have some action.' Ruddell's boys had been several days with no action. I said, 'Sure, I'd be delighted', so we planned one for the next day.

"I was to fly his second section. On a river crossing flight, we would take off and go full bore. We'd fly those planes at 100 per cent power setting until we got out of combat. Engine life was planned for 800 hours and we were getting about 550 or so. Turbine blade cracks were developing. We were running the engines at maximum temperature. You could put these little constrictors in the tailpipe — we called them 'rats'— and you could 'rat 'em up' until they ran at maximum temperature. They were real hot rods. You'd run your drop tanks dry just about the time you got up to the river, and if you didn't have a contact, you weren't supposed to drop your tanks. We skinned 'em every time anyway. On at least four occasions, by the end of the flight I had been to over 50,000 feet. When it

It is fatal to enter any war without the will to win it.
— General Douglas MacArthur

During the Korean War, our task force was operating in concert with other carriers, the *Boxer, Bon Homme Richard, Philippine Sea, Oriskany, Princeton, Valley Forge* and *Essex*. It wasn't unusual for a fatigued pilot, returning from a stressful mission, to land on the wrong ship, in spite of the huge identification number painted on their flight decks. When it happened on our carrier, the *Kearsarge*, the errant aviator had to suffer the humiliation of returning to his own ship with derisive graffiti painted all over his airplane, such as: NO, WE DON'T SERVE BEER HERE EITHER. and ROSES ARE RED, VIOLETS ARE BLUE, THIS IS THE KEARSARGE. WHO ARE YOU?
— Bill Hannan, former US Navy jet engine mechanic

Below: A strike photo showing Navy bomb hits on and near a Korean rail line in November 1951

was empty and I still hadn't pulled the power back from 100 per cent, the airplane would really get up there. The MiG-15 could get right up there too.

"On the flight with Ruddell, we got up there and dropped our tanks. At least the pilots going across the river dropped them. There were big clouds up as we crossed the river. It was early morning and we heard on the radio that there was a fight going on. There were some MiGs flying this day. We came from the sunny side of the clouds to the back side of a big cumulous; there were some black puffs of flak and there were some planes down there. We were half blind from the diminished illumination on the back side of the cloud. It was a confusing situation. We dove down and there was a MiG. Ruddell got it in sight and we dropped from 43,000

feet down to about 15,000, just dropped straight down. He got into shooting position behind the MiG but didn't shoot . . . didn't shoot . . . didn't shoot. I tore past him and blew up the MiG. I had experience at jumping planes and one of the things you did when you came down from extreme altitude (MiGs were frequently found down low when north of the river) was put your armour-glass

defrost on full bore. It would get so hot it was almost painful, but it kept the front windscreen clear. You also tested your guns and your g-suit. Ruddell's windscreen had fogged over. He was sitting there in a kill position and couldn't see properly to shoot. So I went by him and got the MiG. Of course the squadron was abuzz that the Colonel had started crossing the river and gotten aced out of his first kill

Below: A Grumman F9F Panther tangles with the barrier when coming aboard the USS *Essex* (CV-9) operating off Korea in August 1952.

With little more than a hundred hours in the AD Skyraider, I got orders to an AD squadron on the west coast. After I got there I wanted some additional cross-country time so a friend and I flew our Skyraiders up the coast to NAS Alameda near Oakland. While over Los Angeles that night I felt very uneasy because everywhere I looked all I could see was a sea of lights. Over Los Angeles is not the place to lose an engine in a single-engine airplane. There was no place to set down if the engine quit. For several long minutes we flew along over that huge city with me thinking that if the engine quit I would have to ride that free-falling anvil into any unlighted patch I could find. I didn't want to drop an AD into a house. There were no black holes to be seen.
– Paul Ludwig, former US Navy attack pilot

up there. I was a 'MiG killer'. I'd gotten three or four. When I got back, they had all these signs pasted up all over the Dog Flight Quonset hut. One read: MARINE WETBACK STEALS COLONEL'S MIG!

"The 'kill rules' were, if you got seven hits on an enemy aircraft, you would be given a kill. The MiGs didn't torch off at high altitude; they simply would not burn because of the air density. So, incendiary hits would be counted (we had good gun cameras) and if you got seven hits in the enemy's fuselage, the odds were it was dead, and they'd give you a kill. We knew that every third round was an incendiary so, in effect, if you got three incendiary hits showing on the gun camera film, it was considered a dead MiG.

"My first kill was at about 43,000 feet. I had missed a couple of kills before it by not being aggressive or determined enough. I was almost desperate for a MiG kill. I was leader of my flight and I'd screwed up a couple of bounces. My self-esteem and my esteem in the flight was low, and I decided that the next MiG I saw was a dead man and I didn't care where he was.

"The next MiG was part of a gaggle; MiGs as far as you could see. I made a good run on one of them and pulled into firing position, but other MiGs were shooting at me and my wingman, and they were very close. I got some hits on my MiG and he went into a scissor, which was a good tactic. I think the F-86 may have had a better roll rate. I was trying to shoot as he passed through my firing angle. Each time I fired I delayed my turn, so he was gaining on me and drifting back. He almost got behind me and was so close that his plane blanked out the camera frame. I think he realized that I would have crashed into him rather than let him get behind me, and he rolled out and dove. Then I got several more hits on him and he pulled up (he was probably dead at this point). That scissor was the right thing to do; he just shouldn't have broken off. He was getting back to a position where he could have taken the advantage.

"The salvation of the F-86 was that it had good transsonic controls; the MiG's controls were subsonic. In the Sabre, you could readily cruise at about .84 Mach. The MiG had to go into its

Right: Britain's front line fighter in Korea was the Hawker Sea Fury which was operated from Royal Navy carriers such as HMS *Glory*.

uncontrollable range to attack you, and its stick forces were unmanageable. As I recall, the kill ratio between the F-86 and the MiG was eight to one. This was due almost entirely to the flying tail of the Sabre, although it had other superior features. The gun package of the MiG was intended for shooting down bombers like the B-29 and B-50. It carried a 37mm and two 23mm cannons, which was overkill against fighters. Although the F-86 used essentially the same machine guns as a World War II fighter, the rate of fire had been doubled, and it was a very good package against other fighters.

"Down low, where we were out of the transsonic superiority range, we wore a g-suit and they didn't. You can fight defensively when you are blacked out, but you can't fight offensively. If you had enough speed to pull a good 6g turn, you would go 'black' in 20 to 30 degrees of the turn. The MiG pilot couldn't follow you because he was blacked out too. You are still conscious, though you have three to five seconds of vision loss. When you had gone about as far in the turn as you thought you could carry it, you could pop the stick forward and your vision would immediately return. You had already started your roll, and the MiG was right there in front of you, every time, because, not having a g-suit, he had eased off in the turn. His g-tolerance was only half of yours. So he was right there and most probably would overshoot you."

Below: The Chance-Vought Corsair fought its last major action in the Korean War. This example is shown with the late Mark Hanna at the controls. Sadly, forty-year-old Hanna was killed in the crash-landing of a Spanish-built version of the Messerschmitt 109, at an airfield in Spain in September 1999.

SEA CHANGE

THEIR TRAINING STANDARDS are among the highest in the world. They have to be. Naval aviators must be capable of operating safely and efficiently in the most adverse circumstances. When the pilots of a carrier air wing come out to the boat for a refresher carrier qualification (part of the ship's work-up prior to a six-month deployment) they normally practice their landing approaches when the ship is within a reasonably short distance of the beach. They know that if they have to, chances are that they will be able to divert and land safely on a nearby airfield. The practice is good for perfecting their skills and helps build their confidence for doing what is probably the most important and demanding part of the job: bringing their multi-million dollar airplanes, and themselves, safely back aboard the carrier. Once deployed, however, they can no longer rely on that safe proximity to a friendly airfield ashore should they need it. For them, it's the boat or nothing. They need to have worked out any kinks in their approach and landing technique before they can perform safely and confidently in the blue water environment, out of reach of the beach. A pilot who gets too many bolters, or has a tendency to be a bit low in the approach, for example, must cure himself of such habits and achieve consistency in his traps, both day and night, to become truly blue water competent. Key in the final weeks of the pre-deployment work-up is the identification of any such flaws. If these habits cannot be overcome, both mechanically and psychologically, the pilot will be deemed not up to scratch, and unsuitable for operational deployment.

Who are these men and women who fly the warplanes of the Navy? Where do they come from, these pilots and aircrew who carefully don flight suits, g-suits, survival gear, helmets, oxygen masks and more, and shoehorn themselves into tight-fitting cockpits so they can sling-shot from the relative security of their carrier to whatever may await them on their assigned mission? How did

they get to the fleet, these people who, in all weathers and the blackest of nights, must return from their missions and somehow locate their ship, often with no visual cues, and come aboard, no matter what, because there is no other place to land? They are the the best of the best, drawn, like moths to light, to the greatest challenge in aviation.

Jeff Mulkey flies with Helicopter Anti-Submarine Squadron 8 aboard the USS *John C. Stennis*. A 1993 graduate of the United States Naval Academy, he describes the path taken by those who want to fly with the US Navy. A typical American naval aviator comes to NAS Pensacola (Florida), the cradle of naval aviation, from either the Naval Academy or a college ROTC programme, or from Officer Candidate School: "It all starts in Pensacola with API (Aviation Preflight Indoctrination), which lasts for about six weeks. Then you go on to a Primary Flight Training squadron, known as a VT, at either Pensacola or Corpus Christi, Texas. All naval aviators — regardless of whether they become jet pilots, fixed-wing propeller pilots, helicopter pilots, or Naval Flight Officers — start Primary training in fixed-wing, flying the T-34 Charlie, a single-engine turboprop sporty model. Everybody learns to fly fixed-wing and the basic training is all the same no matter what type of aviator you are in the Navy.

"From Primary training it branches out, based on the needs of the Navy in the week that you are graduated. The grades that you receive during that training also affect the Navy's decision about which aircraft community you will be sent to. It will send you to one of several pipelines, the biggest ones being Propeller, in which you would end up flying P3s, C-130s or some other prop type; Helicopter, which can branch out to any of the types of helicopters we fly; and Jets, which could be F-18s, F-14s, S-3s, EA6Bs, etc. If you go into the helicopter pipeline, you stay there at Pensacola and train in the TH-57 up at Whiting Field. The

Your job is to kill people and destroy things in the name of the United States. If you don't want to do that, you don't belong here.

Right: Crewmen of VT-10 aboard the aircraft carrier *Enterprise* (CV-6) in 1944.

pipelines then branch out to Advanced training. For Advanced in Jets, you go on to either Meridian, Mississippi or Kingsville, Texas; for Advanced Propeller, it's Corpus Christi, Texas. At the completion of Advanced training, pilots are 'winged' and officially become naval aviators at that point. Then you go to the Fleet Replacement Squadron (FRS), also known as the RAG or Replacement Air Group. At the FRS or RAG, you learn to fly the particular type of aircraft that you will fly permanently in the fleet. That is where a jet pilot may become an F-18 pilot, or a helicopter pilot may become an HH-60 pilot. There you are sent to your community within the communities at the FRS.

"In terms of elapsed time before you get to the fleet, it depends on the particular pipeline and how smoothly things are running for you, but if all goes well, you can get to the FRS in a year-and-a-half to two years. At completion of the FRS, you head to your first sea tour or fleet squadron. The flow of the pipelines works pretty much the same for all naval aviators, whether helicopter, jet or prop. On average, the total process these days takes you about three years to get to your first fleet squadron — a pretty long time. But they are making an effort to weed out some of the 'pools' that have built up in the past, and the demand right now, in the year 2000, is certainly high for pilots in the Navy, so they are beginning to move more quickly through the pipelines. The primary reason for the increased demand is the number of mid-grade pilots that are electing to leave . . . that are not staying on in the service. The Navy is losing a lot of the more senior folks; the mid-career officers who have finished their initial commitments and have opted to get out and pursue other interests, whether they be commercial aviation or some other type of job. The senior leadership is still there. The new guys are still coming in, but the middle is eroding a bit. It's the same across the board . . . fixed-wing,

Right, centre and far right: Shannon Callahan makes her way through Naval Flight Officer training at NAS Pensacola in January 2000. Ejection-seat training was followed by "the slide for life" water survival exercise. Below: Winged and rewarded with the rank of Lieutenant Junior Grade, Shannon stands before an EA-6B Prowler, the aircraft in which she will fly with the fleet as an Electronics Warfare Officer.

rotary and jets. I'm one of those mid-tour guys. I'm just finishing my first sea tour and then I'm going on my first shore tour. I'll have the option of getting out in about three years. I'm keeping all my options open.

"Everyone in naval aviation is aware of their obligation. No one enters into it without realizing how much time they're gonna owe. But life changes. Most people are not married and do not have children when they start flight training. If you do get married and have children, suddenly you are five years into it and your outlook may be quite a bit different from what it was when you were single. It's talked about — separation — but until you have actually done it, until you have gone through a carrier work-up cycle, until you have been away from home for six months or more and gone through a cruise and experienced what it's like, only then can you know. Everyone thinks at the outset that they are ready to accept the life and the obligation, but they don't really *know* until they've been there. The Navy sets the commitment that both it and the individual agree to and accept. It's a contract and all sides understand that. You know what the Navy expects and everyone honours that. Certain people deal with it better than others; some enjoy it more than others, but the contract is always honoured, whether for personal reasons or because the Navy says you will or out of a sense of honour and obligation or because of monetary reasons. A whole lot of money is spent training a person to do what we do and the Navy has every right to expect to get our services for a number of years, for training us to do this."

Dale Dean, Director of the Aviation Training School at NAS Pensacola: "We give you a test to see how smart you are and how good you are at problem-solving. If I were to define the perfect person for us, it would be someone who scored well on the test and is about 5 feet 8 inches tall.

That's important because we have anthropometric concerns. When you are sitting in the aircraft, you have to be able to see over the glare shield, to reach all the switches, to fully throw the controls in out-of-control flight, maybe negative g, and you have to be able to reach the rudder pedals, and, if you have to eject, not tear your legs off. If you are 5 feet 2 inches tall and have little T-Rex arms, you're probably not going to qualify. If you are 6 feet 7 inches tall and your legs go under the glare shield, you are probably not going to qualify, at least not for jets. We want that academic profile and we're looking for a fairly athletic, medium-build person. And then there is the part that is hard to define . . . motivation. You have to really *want* to be here. You're not just coming to get your ticket punched,

Our fam flights in the T-34 are very similar to the pilots' fam flights. We actually get stick time there, to build our situational awareness so we know what the pilot needs from us. We have to know all the emergency procedures. If the pilot get disoriented we talk him into getting straight and level, so you have to know a lot about aviation.
– Shannon Callahan, Naval Flight Officer student, NAS Pensacola

"We have to do the dunker every two years. As aircrew, you have to do six runs. The first one, you go to Yeovilton where the dunker is. They strap you in the seat in the front. Then they down into the water to the bottom of the pool and, when the thing has come to rest on the bottom, you make your escape. Then they do a series of runs that get more and more difficult. In the final one, they don't even really give you a chance to strap in. It's a sort of uncontrolled situation where they turn the lights off around the pool so it's pitch black. Then they drop the dunker into the water and spin it around a few times so you end up upside down. They also fix it that some of the exits that you would normally use to get out of are blocked; the windows have buckled or whatever. You actually have to use the little oxygen supply which is almost like a little aqualung, that we carry on our survival vests. You practice getting that into your mouth, controlling your breathing using it, and then finding your way through the aircraft to an alternative exit. We also do Culdrose every six months, where we jump in the pool and inflate our life jacket, get in our little personal survival pack which we clip to our life vest and sit on throughout the whole of the sortie and which, in the event of a ditching, should come with us. Abandon Aircraft drills . . . we do wet-winching so we'll be thrown into the sea and picked up by a search-and-rescue helicopter or a helicopter from our squadron. That's done

spend a couple of years flying here and then go on to the airlines. How badly do you want to be here? Are you willing to gut it out and put up with the inconveniences to get to fly?

"So, it's a three-part requirement. You have to be able to do the job, fit in the plane, and really want to be here. The people who do best as Naval Flight Officers are not necessarily the ones who have engineering degrees; it's the ones who really want to fly, they just *want to fly*. They are really smart, have good eyesight, good hand-eye coordination, a quick mind, able to quickly grasp detail. In general, most successful naval pilots or NFOs are successful naval officers as well. They do pretty well with their ground jobs, are multi-faceted and able to handle more than one detail at a time. A lot of it is attitude. You have to want to be here. You have to enjoy it. You have to want to be in charge, to take control of the situation. That's where success comes from."

What sort of person flies the F/A-18 Hornet front line fighter in the US Navy today? One example is David Tarry of VFA-147 on the *Stennis*: "I've been flying the F-18 for about fourteen months. Ten months getting through the initial training and four months with the squadron. The primary purpose of the airplane is to cover both the fighter and attack roles. We have different master modes for our computer that let us fight our way in, drop our ordnance and then fight our way out, without dedicated fighter support. The Navy is moving towards a multi-role airplane and away from the dedicated fighter and the dedicated attack plane. The F-14 will be phased out and the Super Hornet will be our next line, followed eventually by the Joint Strike Fighter (JSF). The Super Hornet will also take over tanking, which the A-6 used to do, and will be the future of naval aviation until we can get the JSF on line. We're supposed to start training pilots for the Super Hornet next year [2001]. It will have a dual-seat

configuration. Right now, the Marines fly a dual-seat variant of the F-18 — land-based — for doing forward air control and high-task-loading issues.

"When we launch from the cat, the F-18 pilot grabs the 'towel rack', a grab handle on the canopy frame, prior to and during the launch. The aircraft rotates on its own with the trim that we set, but right after you come off the front end, you grab the stick immediately. They just don't want you to put in any stick control because it would disturb the trim. You can do the same taking off at the field. It's pretty much hands-off. As you get the air speed and wind over the wing, it just naturally rotates and takes off. It's a very user-friendly plane. The computer systems we have allow us to go from a two-seat cockpit, like the F-14 or the A-6, dedicated fighter or dedicated attack, to where we can do all of that with a single-seat plane because it is so user-friendly, with the hands-on throttle and stick system (HOTAS). You can run your computer and all your displays, and almost never take your hands off the stick and throttle. It takes most of the work out of flying and actually lets you concentrate on working the weapons systems and flying the plane as a weapon. Higher level guys who have at least a few hundred hours in the plane are very comfortable working their way in and working their way out, and not having a backseater to help them. The computer is really your backseater, taking care of a lot of the administrative tasks.

"To land an F-18 on a carrier, you are coming in at somewhere between 130 and 140 knots and trying to hit that little piece of ground and stop that quick. Lots of my friends can't believe I do it. They can't see how you can do something that precise. But really, with the training you have, you just start off and everything is baby steps. They never have you take a big step. You start out in a small turboprop, just learning basic flying and how to land at the field. You go to an Intermediate jet trainer where you start working with higher speed manoeuvres, but still just working at the field. You get into Advanced training and start learning more tactical things. You start doing some bombing, some air-to-air fighting. Then you do a long period of field carrier qualifications, using the lens to land at the field. They work you up and work you up, and then you go to the boat in Advanced during the day. Then you go to the FRS, the RAG, where you learn how to fly the F-18, the F-14, S-3, or whatever you happen to be flying. You learn how to do the bombing, the air-to-air fighting, and then you do day and night field carrier work and, eventually, carrier qualification. But everything is done in baby steps. By the time you get to the fleet, it's not an unfamiliar environment. That's a big step. Seeing the sight picture; looking down from 16,000 feet while holding overhead and thinking, 'I'm gonna land on that postage stamp down there', sure gets your heart rate going. But they baby-step you along so much that it's really not a huge leap of faith to be able to do it.

"When you learn to land at night on a carrier, it's basically the same pattern that you fly during the day. So when you make the transition to night, you're basically doing the same thing you do in the day. You just don't have the visual reference, but if you do everything at night the way you are supposed to do it during the day, you should roll out in exactly the same spot. Again, it's a little step. Take away the visual reference. But you do it enough at the field, and you start flying extended patterns, or holding in a marshal stack and then flying straight in to pick up the Instrument Landing System (ILS) or the Automatic Carrier Landing System (ACLS). I find that night landing is actually a little easier, because you are doing a straight-in. You aren't doing the approach, hitting specific numbers and having to roll out. Rolling out, making the transition from the turn to wings level, is a pretty big power correction. With the small, stubby wing

that we have, you're pretty high up on the power in that approach turn, so when you roll out you get all that lift back under your wings. You've got to come quite a bit off the power, keep the rate of descent, and then get back on the power so you can find the middle ground again. You're really fighting to find the middle happy place. But if you're coming in on a straight-in approach, you are three miles out and making the transition from straight and level flight to about three degrees glide slope down, a little power off, a little power back on, establish yourself on glide slope, and then, for three miles you're just fine-tuning. By the time you are flying the ball, it's just little corrections and maybe a little power on as you fly through the burble, depending on how the winds are across the deck.

"Once you get used to flying at night, and get over the mental aspect, which is really what the problem with night is . . . it's more of a mental thing . . . that you don't have the horizon and the visual reference, but once you get over that it's really not bad at all. You use the same rules that the Landing Signal Officers (LSO) teach us all the time. If you're a little high, lead it and step it down. If you're low, get it back up where it's supposed to be. There is no life below the datums. You don't want to be down there, so get it back up. Ball flying is ball flying, whether it's dark or light out. You do so much of it at the field that it becomes more muscle-memory than anything. 'OK, the ball is down. I've got to get it back up.' It becomes subconscious. They talk about the zen of ball flying. It's an art, it's a religion to have that touch, that feel. I don't know that I have it yet. I'm still working on it.

"The F-18 is fun to fly. It's the greatest job in the world. Every day I fly, I can't believe that I'm doing it. I can't believe that anybody would let somebody do it. We're no different from anybody else; there's nothing that sets us apart from everybody else except, maybe, that we have the eyesight to get in.

every two years as well. We have a naval swimming test which is to swim about fifty metres in overalls. You have to do that at Dartmouth or in your basic training as soon as you join the Navy. If you don't pass, then you have to go into remedial swimming lessons."

The air fleet of an enemy will never get within striking distance of our coast as long as our aircraft carriers are able to carry the preponderance of air power to sea.
– Rear Admiral W.A. Moffett, Chief of the US Bureau of Aeronautics, October 1922

He got into the gene pool while the lifeguard wasn't watching.

Above: Royal Navy helicopter pilot Claire Donegan. Top left: Pensacola navigation class in 2000. Top right: Keri Lynn Schubert, the first female US Marine Corps F/A-18 Naval Flight Officer. Above far right: Escape training in the helo dunker on the NAS Pensacola Naval Flight Officer course.

That's the big thing with carrier aviation. It keeps out a huge percentage of people who would love to do this and are more than smart enough to be able to go through the training and do it. But the Navy has such rigorous eyesight standards, mostly because of night landing. You have to have really good eyesight. The Hornet is a great airplane; a great ball-flying airplane. It's easy to put the ball where you want it, easier than the F-14 or the Prowler whose systems are more manual and take a lot more work to fly. The F-18 has a great system that helps us to be safe and get aboard safely. Taking off the front and landing on the back is more admin than anything else, and you just want to be able to do it safely. The LSOs are always pushing: 'Just be safe.' You don't want to be low on approach because it's not safe. If you are high you can always go around and try again. If you do that four times and you have to tank, we have tankers. Go get some more gas, but always be safe. Never drag the ball in low just to get aboard, taking the risk of settling right at the back end of the boat. Our bosses tell us that our job is to go downtown and drop bombs, or fight our way, or defend the strike package that's going in. Getting back on the boat is just admin, but it's one of the prime spots where you can get in trouble."

"Usually, a student is kept free of trouble by instructors, and the student tries to keep himself out of trouble because he wants to succeed alongside his peers. But stupidity appears to many of us at the very time we are happiest. In 1954, during Primary Training in Pensacola after I had soloed, I thought I had mastered the SNJ [trainer], and one day over Perdido Bay I was practising aerobatics for my next check ride. I did a slow roll and hadn't got the nose high enough on the entry. When I was upside down I let the nose fall through, and, rather than roll out, I split—essed. I blacked out from g during the pull-out and awakened to see the Indicated Air Speed needle exceeding the red line, and expected to see the wings tear off, but the airplane held together. It shook me up but I got over it."
– Paul Ludwig, former US Navy attack pilot

Royal Navy pilot Danny Stembridge flies FA 2 Sea Harriers with 801 Squadron from HMS *Illustrious*: "I went through flying training to fly the Lynx maritime attack helicopter. I had a fantastic time doing it and part of me actually still misses it. It wasn't as satisfying as flying a fast jet, but I think it was more enjoyable. When you are flying a fighter, there is very little time to think, 'Wow, this is fun.'

That's more in retrospect. Now I think that the satisfaction I get from flying Sea Harriers outweighs the fun I had flying helicopters. The Sea Harrier is very demanding. It's not hugely ergonomic inside, and you are working hard because of that. The air defence environment is a busy one, and you're on your own. Then you have to come back and land on a ship in a VSTOL fighter. Bringing the Sea Harrier alongside, with its lack of systems to help you and its inherent lack of thrust, and therefore the amount of fuel that you require to come aboard . . . it takes a lot of brain power and piloting ability. You're working from the minute you start preparing for a sortie 'til the minute you finish debriefing it, and afterwards, if it's gone right, it's immensely satisfying. It's a job that has to be done, when you're airborne, whether you are training or not, and the satisfaction comes afterward.

"If you mishandle the Sea Harrier in any way, particularly in the VSTOL regime, it will bite you hard. It will really punish you if you make a basic error in VSTOL regime. But the training is pretty intense and pretty long as well. Because of my rotary flying background, I had a slightly different pipeline, but the Sea Harrier training is a very long course and a lot of it is spent in learning to tame the plane in the VSTOL regime, just flying it. With other fighters, once you've learned, they are similar to other types, but with the Harrier, you are always learning how to fly it and how to operate it in the role for which it was designed.

"Coming from flying helicopters, I have found it easier and a lot more comfortable around the ship than other people do, because of my experience of about 1200 deck landings on very small ships in a helicopter, many of them at night. Ship-borne operations are unique, and to have that to fall back on when everything is looking horrible . . . you're coming down the glide slope, you're in cloud and wondering if the ship is actually gonna be there at the end . . . I'm used to seeing the ship pop out of

the gloop at the bottom. Other people may not be. I also think that the hovering experience has helped me immensely.

"I'm looking forward to my first deployment to the Gulf flying Sea Harriers [in January 2000]. I've been to the Gulf twice, for six months, so the area is not new to me, but the mission I will be carrying out will be totally different and, professionally, I'm looking forward to it. I assume we are going to be doing enforcement of the No-Fly Zone most of the time, as well as OCA (Offensive Counter Air) . . . escorting bombers into a target zone, and DCA (Defensive Counter Air), or sitting on CAP (Combat Air Patrol). We will be there and I assume American ground and carrier-based aircraft, along with those of other nations, will be there as well.

"As for the danger involved in the flying, it doesn't keep me awake at night, nor does going to the Gulf. In the past when I've done what could be construed as scary things, that hasn't kept me awake. Maybe in six months my view will have changed. But I think if you're the kind of person to do this, then it's very unlikely to worry you. Otherwise you wouldn't be doing it."

Shannon Callahan is the granddaughter of a US Navy enlisted man who served on the *Gambier Bay*, a

Above: NFO student Scott Whelpley at Pensacola.

159

Above: Navy ladies at
NAS Pensacola, believed
to be *circa* 1918.

small carrier that was sunk in the Battle of Leyte
Gulf during World War II. In March 2000 she
completed the Naval Flight Officer course at NAS
Pensacola and was selected for the EA-6B Prowler
community: "When I was in high school, I took a
trip to the Air Force Academy and fell in love with
the place. I definitely wanted to go there. But the
Air Force officers I talked to convinced me that, to
be in the Air Force one should really be a pilot, and
I didn't think I could be a pilot because I didn't
have 20/20 vision. I looked at the other service
academies and decided on the Naval Academy,
where I learned that even without 20/20, (I have
20/40) I could qualify to be a Naval Flight Officer. It
is ironic that I actually could have been a pilot in
the Air Force, because their eyesight requirements
are lower than those of the Navy, but I didn't know
that at the time. Fortunately I found my calling
because I really love carrier aviation. I went out on
a carrier on a midshipman cruise one summer
while at the Academy, and was enamoured with
the whole thing. I got five cat-shots in a Prowler
and after that there was no going back. I graduated
from the Academy in '98, got my Masters degree in
National Security, and showed up at flight school
in Pensacola in '99.

"My first choice would be RIO on an F-14; second
choice would be Electronic Counter-measures
Officer on an EA-6B Prowler. The Prowlers will
probably be replaced by the F/A-18G; they call it
the Growler. So it seems like I will end up in the
Super Hornet if I stay in, which I really want to do.
I'll be happy with whatever I get. Every aviator loves
their platform. Nobody ever regrets the direction
they went or their choice. Everyone is positive and
always trying to sell you on their platform.

"Every event, every training flight, is graded. They
keep track of how each student is doing with what
they call Navy Standard Scores (NSS). The scoring
system for flights is: Above Average, Average,
Below Average and Unsatisfactory, which is a
'Down'. Two Belows equals one Down. Your NSS

reflects how many Aboves and Belows you have, compared to the 200 to 300 students who went just before you, so it's all comparative. This is because the instructors come and go, and thus the grading changes a little bit; it's kind of fluid. To keep track of where I am in relation to my class mates, it's just a matter of counting up those NSS grades, and some people do that. That's not something that I obsess about. You'll drive yourself nuts if you keep comparing yourself to your buddies. The thing to do is just to go into every flight and give it your best, 100 per cent. I have faith in my instructors. I think they are pretty good and pretty fair, and I know it's all gonna work out the way it should because these people have been in aviation for years. I trust them with the grades. I'm sure people on the outside probably think it's pretty cut-throat, but it's not. I've never known anyone who screwed a buddy to get ahead in this programme. Everybody always helps everybody else. You *have* to cooperate to get winged. That's hard enough, and there is a huge attrition rate, even in Pre-flight, so you really can't make it without your buddies. The workload is so intense that you have to split up the work. I haven't met any Lone Rangers. You just can't do it. I've worked in groups from the beginning. The instructors encourage it. I think it's a necessity. You have to, to get it all done. And you rely on your buddies to cheer you up, to keep you stable and centred. It can be pretty stressful.

"There are no particular problems in being female in the programme. There are still not that many of us, so, naturally, most of my friends are guys. There hasn't been anything that has really got in my way. I think that if anybody does have prejudices and stereotypes, they brought them with them from the civilian world into the Navy . . . the way they were raised or the area of the country they come from. I don't think the Navy breeds that at all. Most people just want to know, can you do your job? Are you safe to fly with? The percentage of women in the programme is small, especially when you get to this stage. Right now, I think there are six or seven female students in the squadron, out of over a hundred students. There are few of us, markedly fewer of us, but at the same time we are not such an oddity anymore. There are not many female instructors. I've flown with one since I started, and that's from Primary on. I work in a pretty much all-male environment but I'm used to it. The Academy trained me for that. It would have been a bigger adjustment had I come from a civilian school where the ratio was 50/50, so I was kind of lucky in that respect. The military is kind of inherently masculine, and a lot of women compensate by being more masculine. That's not necessary anymore. I think the situation has changed. You can just be yourself. Everything feminine isn't bad any more, and it doesn't necessarily mean you are weak to be feminine. Again, people are interested in the bottom line. Can you fly? Can you navigate to the target? Can you get to the target on time? Are you safe? Do you know your emergency procedures? Can you make good judgement calls? Do you do thorough pre-flight planning? All that stuff is pretty gender–neutral.

"To get through this programme you really have to love it. That's what I tell people who are just starting. If you don't love it, don't bother, 'cause it only gets harder and more demanding. Your attitude has to be: 'I *get* to do this.' Not, I *have* to do this. I'm really lucky to be doing this. I saw a lot of my buddies get attrited for medical problems or because they didn't have the grades for Aviation. I've been really fortunate to be in this programme and I'm not gonna make myself miserable just because it's hard. You have to earn something to appreciate it. I knew a long time ago that I wanted to be in the Navy for a long-term career. If you love your job, it's the biggest blessing in life. I would do this if they paid me in sand. I'd like to command a carrier one of these days. Right now, I'm just trying to get winged and fly as much as I possibly can.

Without seeing a viable threat, the American people are not going to want to spend more tax money on a big military. For what reason? There's no big war going on right now [January 2000]. The fact that we've got hot spots all over the world; that we've got troops committed, is lost on the average tax payer. When is the last time you saw any TV news from Bosnia or the Gulf, or any of those places? We've still got troops there and we've still got aircraft flying right overhead. We have a large presence in the Arabian Gulf. We've got an aircraft carrier there all the time. People don't realize that. As far as they know, there is peace all over the world, and there's not. But they are not going to increase the budget. You don't build a huge military to go fight little regional civil wars, even though they might escalate. We won Desert Storm and suffered from the "peace dividend". From World War II on, we had the threat of communism. Where is the threat now? I don't want to see another war, but I don't know what the military is gonna do to recover from this [down-sizing]. I don't know what the fix will be. Maybe there is none. Maybe we just have to live within our means and find a creative way of recruiting the best and brightest, challenging them, taking care of them and making them feel wanted in this job. Then maybe we'll be OK. The Joint Chiefs of Staff went before Congress again about a year ago and briefed them. From what I've seen, their conclusion was: "We can no longer do more with less; we can only do less with less.
continued on page 162

161

"The Naval Academy is a great place. It had a big part in making me who I am today. I'll always be grateful and proud that I went there."

Recent Pensacola graduate Scott Whelpley is also headed for the Prowler community. Scott made it to the NFO program by an unusual route: "For nearly six years I was an enlisted Aviation Electronics Technician in the Navy. I applied for Officer Candidate School for aviation and was turned down. I applied again and was accepted, but for the Supply corps. I was told that I was too old for naval aviation. After the tenth week at OCS I asked if I could switch to aviation. 'You're too old and there's nothing we can do about it.' I was thirty-one at the time. So I was commissioned a Supply Corps officer. Then I met a lieutenant commander who thought that what I had been told might not be true. He himself had been thirty-one when he went through the NFO programme. He made some calls for me and, after about four months and more rejections due to my age, I was finally accepted.

"I got a month's advance pay, threw all my stuff in a U-Haul and was on the road to Pensacola. I'm older than most. I just turned thirty-three. A lot of these guys out of the Academy are twenty-three. I really admire them. They are so smart and grasp this stuff so quickly. I'm a little more stressed and concerned because I feel like this is my last hurrah, so I really need to get through the programme. I'm having a good time but I've put a lot of pressure on myself because this will be the greatest thing that I've ever accomplished. I study all the time. I have to. I know the physiology guys would throw a fit if they knew, but there are times when I'm up 'til one and then get up at five, because I have to. It's the only way I'll get through it. It's been ten years since I was in college.

"Everybody here wants to be the number one flight officer in the programme. I know I'm somewhere in the middle of the pack but that never stops any of us from competing and trying to be better and better. We are tested and graded every single day. We have to know all the systems of the aircraft, all the procedures, all the checklists, all the emergency procedures . . . every single day something is added on top of what we just learned. But I'm so happy to be here. I'm very lucky. Back home everybody is like, 'Wow, Scott, you're gonna be flyin' a jet.' If only they knew. It's just me, *me*, the same guy you used to run around with and get in trouble with. If I can do this, anybody can do this. Obviously you have to want it badly enough. It's a Big Boy programme here. You are responsible for knowing your stuff. Go ahead. Go home and do whatever you want to do, but you'd better know your stuff when you come back. It's taught me discipline. Being 'enlisted' really humbled me and I think I needed that. It was what I needed prior to doing this. I think it has made me a better officer.

"Yesterday was my fourth hop. Today will be my fifth and after that I have seven left. I'm not a pessimist or an optimist; I'm a realist. Anything can happen, so I'm just hoping. Hold on to your hat and don't let go. Eight more events and that's it, I get my wings. As soon as you are done with your last flight they come out and wet you down with champagne. They rip off your name tag and put on your 'soft wings', and you get a punch in the chest, which is politically incorrect. After that we have a giant ceremony at the Naval Aviation Museum with the families invited. After we get winged, we are selected. I'm probably going to get S-3s or Prowlers. My first choice is Prowlers. I've been in the Navy for six years and I'll owe them six after this. I really enjoy the Navy. It's funny, I miss going out to sea. That might seem crazy but I really do. It's one of the few things in this world that is still an adventure. You never know what's gonna happen. I was on a small aircraft carrier, the USS *New Orleans*, in 1997 when we took a 45° roll during Typhoon Justin off the coast of New Guinea. That was pretty scary. We almost lost it. I've never been so frightened in my life. Everybody

I'm in the Naval Aviation Choir. There are only two organizations that are ambassadors for naval aviation: The Blue Angels and us. Ours is the only all-officer choir in the military that I know of, and it's a good mix of people, students and aviators. We travel around the country and we do a lot of local gigs. We sang at the Christmas open house at the White House.
– Shannon Callahan, Naval Flight Officer student, NAS Pensacola

was surprised that the island didn't break off. The ship had some really heavy damage. That is real adventure; it's dangerous, but there is nothing in the world that I would rather be doing. No matter all the grey hair, I plan to stay in as long as they will keep me. I'll get to the highest rank I possibly can. It's all about being proficient. I'll do whatever it takes. I'm willing to learn, and wherever the Navy takes me, I'm ready to go.

"Grades. Sometimes you get bad grades. Sometimes you have a bad flight. It's humbling and sometimes you feel you're not worthy. I had a great hop on Friday, Above Average, and I felt like I was *the* best flight officer. Yesterday I had an average hop and got an Average grade. Driving home, I was kicking myself — what did I do wrong? I suck. I'm horrible. It can go from one extreme to the other. You just push it all away and go back the next day.

"The days and weeks have flown. I'm almost afraid to look up and see how much more of the hill I

have to conquer. I like looking back down behind me and thinking, 'Hey, I've come this far. I'll just keep my head down and take one day at a time, one step at a time, and just keep chuggin' away.' "

Another who came late to flying is Nick Walker of 801 Squadron, Royal Navy. Nick was a ship driver until July 1994, when he entered flying training: "I hit the front line on Sea Harriers in July '99 and have about 150 hours on the aeroplane now [in October 1999]. Our orientation for going to the Gulf in January has really been going on all through our flying training, because the Gulf thing has been going on since 1990. The Sea Harrier training squadron is given a weekly brief by an Intelligence officer who will often cover the Gulf or Bosnia. Before deploying to the Gulf, we will go through a series of formal briefings and the rest of the squadron will get involved in station-level briefings by Intelligence officers. The week before we sail we'll get a threat update and an analysis

Above left: T-39 and T-34 instructors at NAS Pensacola, Florida in January 2000.

Above: Base facilities and aircraft at NAS Pensacola, Florida, believed to be *circa* 1930.

update. Once we are on board, the briefs will continue. We'll be well prepared for the threat we are possibly facing out there.

"It seems strange to say, but I hope I never have to go and do my job for real, because if you do go and do the job that we train for for real, it means that we are at war and diplomacy has failed. We will be patrolling the No-Fly Zones set up over Iraq and it's quite possible that we'll have contact with Iraqi aircraft. We'll be flying in conjunction with American aircraft which provide the bulk of the numbers on the combat air patrols there, with help from the Sea Harriers and from various other countries as well. We'll be working very closely with the Americans.

"The Sea Harrier is unique. It's not the fastest or the most modern aeroplane. It's got a great radar, a great weapon, which makes it very formidable as an air defence fighter. And it's got these nozzles which make it unique. Learning to fly it was . . . very different. You get used to learning how to fly

aircraft faster and faster and lower and lower, and then you get to the Sea Harrier and they teach you how to stop in flight. From about 100 knots downward, its wing is not very efficient and produces very little lift, so once through 100 knots in a decelerating phase, you have to progressively put more and more power on to keep the aircraft in the air. Coordination. It certainly concentrates the mind, recovering to the ship. Going into a big airfield is fine. Half the time we land conventionally, going forward at speed. But on the boat it's vertical landings every time; recovering to a moving airfield that is going up and down as well as forward, is a challenge.

"Coming back to the ship is very well controlled. We have a certain time to be landing on deck. In the initial brief we will know what time we are landing on board; it's called a Charlie time and it is given to us by the ship to fit in with the rest of the ship's programmes. Once airborne we are managing our fuel, both to achieve the mission aims, do our training or whatever, and also so that we can arrive back in the vicinity of the ship with the correct amount of fuel, so we are not too heavy for the vertical landing. You have to be quite critical and careful about how you manage your fuel. If you are going to start dumping it, which you have to do on most sorties to make your weight back at the boat, you need to dump down at the correct time, generally about fifteen minutes before your recovery so you can properly slot into the boat.

"Assuming it's a nice day, it's then up to us to come back to the boat on our own. We might get some assistance from the air directors in the Ops room who can pick us up on radar and give us steers back to the overhead. We then set up a holding pattern 1000 feet over the ship. The pattern is about two-and-a-half minutes long. You go round in a circle above the ship and, two-and-a-half minutes prior to your landing, or Charlie time, you fly past the ship at 600 feet and slightly on the starboard side of it. That's called the slot, and once you have slotted, you then turn in front of the boat so that you are effectively on a cross-wind leg and you roll out going down wind . . . the opposite way to the boat. As you approach the back of the boat, with all your checks complete and your landing gear down, you tip into the finals turn and it's around that turn that you start to wash off your speed by using some of the nozzles. You take about 40° of nozzle around the finals turn, maintaining it flat initially and then using about 90 per cent power, monitoring the engine temperature so you don't trip off the water injection system. You use the nozzles around the final turn. You are playing the nozzles to keep your speed and your angle of attack under control around the turn. As you approach, you roll out behind the boat and start in towards it. You then judge when to take the hover stop, to move the nozzles all the way downward to allow the aircraft to hover. Obviously, this means that there is no

thrust going out the back of the aircraft, so it starts to slow down a bit more rapidly. The idea is to arrive alongside the boat in a steady hover next to the landing area you have been allocated for that particular landing, at about 80 feet or so above the sea level and just off to the left side of the boat so you are not quite over the deck but fairly close to it. Once you have steadied yourself into a nice position and you're happy with your references next to the boat, you do a transition across by moving sideways about 30 feet to line up just over the mark that you have picked. You reduce power just a bit and the aircraft starts to sink. It's quite a firm landing because on a wet or pitching deck, if the wheels don't make a good contact, it would be easy to slide about.

"You think about the danger, although not all the time. You're very aware that the Sea Harrier is a difficult machine to fly; that if you don't keep on top of it, if you don't concentrate during decelerations and accelerations into and out of the hover, it can go wrong very quickly. There is a higher chance in this aeroplane that you may have an accident or crash or have to eject, or even be killed. It's something that you do think about periodically. I don't think about it when I strap into an aeroplane, when I go flying. The training you get is very good, and there is no way they are going to send someone out to fly the aeroplane if they don't think they can do it. You think about the danger when you have aircraft systems failures or when you have a problem with the aircraft . . . when you are just that little step closer to maybe having to eject. Fortunately they are rare. When something happens to someone you know, that brings it home to you. Yes, it is a dangerous occupation, and yes, you do have to stay on top of the airplane at all times, but I don't think it would ever stop me doing it. I don't think it would ever make me so scared that I wouldn't want to do it."

Jason Phillips is an Observer Instructor and

The F-14 was the premier Navy fighter when I received orders to fly it. I'll never forget the day I got my Navy "wings of gold", 30 December 1982. I was the only one to get orders to fly the F-14 and my training duty station was to be NAS Miramar in San Diego. My father gave me a Rolex with the date and Navy wings engraved on the back. I realized later what an expensive gift that was (he had a $20 Timex) and it remains my most precious possession and will be passed on to my son.
– Frank Furbish, former US Navy fighter pilot

Warfare Instructor in Sea King helicopters with 820 Squadron, Royal Navy aboard HMS *Illustrious*: "Our primary role on the squadron is to hunt submarines. We do that using sonar and there are two sonars on board. The active sonar is like a big microphone, a long bit of electric string that we lower into the water. We press a button and it sends out a pulse of noise. If there is anything out there, the pulse bounces off it, comes back and is displayed on the computer's green waterfall display. Because we put out a 'ping' when using active sonar, that lets the submariner know that we are there. Ideally we like to sneak up on him, to give our weapon the best chance. We also use a passive sonar: a smaller microphone on a thinner bit of electric string. It goes down in the water and just listens. It floats to the surface and has a little radio antenna. Everything it hears is uplinked to the aircraft and displayed on the same screens. No noise, no ping, it just listens. A trained aircrewman with super-sensitive hearing will pick out the contact and pass the information across to the observer who is the tactical coordinator. We can be the tactical nerve centre when we get a contact with a submarine. I would then take over as the scene-of-action commander. No matter the seniority of the ships in the area, we can tell them to move out of the way while we deal with the submarine. We control the weapons and carry up to four [torpedos] on board. The weapon we use is a Stingray, a highly intelligent weapon. Another bit of equipment we have is the MAD or magnetic anomoly detector, which detects magnetic disruptions in the earth's field. It's a big bar magnet and if we fly over something like a submarine, a big lump of metal, it causes a disturbance. We have a sensor which a crewman operates, and the contact is also painted on an old-fashioned chart recorder. If we think there is a target down there, we can fly over it just to corroborate our information.

"The Navy is now gearing up for the Merlin,

166

which is gonna be our new aircraft. The Merlin is at Culdrose now, going through intensive flying trials and is due to go front line in September 2001 in *Ark Royal*."

Michelle Vorce graduated from the US Naval Academy in May 1997: "As I went through the Academy, I wanted to go into the Medical Corps. Aviation wasn't my first choice. I'm unusual in that, for about 95 per cent of the people here at Pensacola/Whiting Field, aviation *was* their first choice. I had been selected for med school and by the time we had to turn in our final selection sheets, I hadn't heard back from the medical schools where I had been interviewed, and I had to make a selection. I had done a one-month cruise on the carrier *Kitty Hawk* and was able to fly all sorts of aircraft, so I decided to give aviation a shot and see how I liked it. I chose it as my first choice and got it. I went through API at Pensacola and then did my Primary at Whiting. After four-and-a-half months I selected Jets and went to Kingsville, and was there for six months before they realized that what they called my anthropometrics, my height measurement, didn't fit the fleet. I was sent back to Whiting to fly helicopters and was winged on 17 December 1999. I'm going to San Diego to the Sea Hawk/Bravo community.

"I went through the Academy from '93 to '97, which was after all of the problems with sexual harassment. I never once experienced a problem. I got high leadership positions without a problem and I think that I was generally respected by both male and female counterparts. You are always surrounded by males but I never experienced a problem. I fit in real well and was accepted and respected. In training it has been on a par or even better, especially in the helicopter community. We [females] are still a definite minority, especially in this field, but it's getting better. The instructors are very respectful, never surprised or shocked at our ability or competency. If you study hard and

perform well, and you show that you've earned the right to be there, just like your male counterpart, you are accepted right off."

"In the Navy Pre-flight school that I went through, half of the time was spent on such relevant subjects as aircraft recognition, theory of flight, navigation, and naval traditions and customs. The other half was devoted to body contact athletics. Football was rough enough. We played basketball with ten men on a side, and the referee was there solely to break up the fist fights. We played water polo in the deep end of the pool and the instructor cracked the knuckles of anyone caught hanging on to the side. Pushball was another form of mayhem designed to inflict bodily harm on the other guy. In hand-to-hand combat we learned how to knife-fight and how to break a man's neck, go for the jugular, break an arm, dislocate a shoulder, apply a knee to the groin, and so on.

"Physical fitness training, where competitiveness was stretched one step further to combativeness, was re-emphasized each step of the way in the flight training programme, which lasted about ten

Left: FA 2 Sea Harrier pilot Pete Wilson of No. 801 Squadron, Royal Navy in the cockpit of his aircraft. Above: A tilt-rotor MV-22B Osprey, which is scheduled to begin service aboard US Navy assault carriers in 2002 with the US Marines, providing them with a much faster and deeper inland penetration capability than was previously afforded by helicopters.

months at that time [1942]. I think it is fair to say that we came out of the cadet programme with a certain killer instinct and an aggressive approach to survival. When it came time for me to go into combat against the Japanese, the issue was uncomplicated. They were out to kill me and I was determined to do unto them before they did unto me."
– Ed Copeland, former US Navy fighter pilot

Once told by his Navy flying instructor, "You will *never* solo. You are the *dumbest* cadet I have ever laid eyes on", US Marine night-fighter ace Bruce Porter became one of the great fighter pilots of World War II. "As I stood watching my first night carrier ops, the feeling that crept over me was as eerie as any I had ever had. All I could hear was two high-performance engines. All I could see were two sets of landing lights. There was absolutely no moon to light the wake or give a hint of the ship's whereabouts. There was a slight breeze over the bow.

"Suddenly, one set of lights was over the wake. The approach was textbook perfect, right up the wake at precisely the correct speed, attitude and altitude. I heard the engine drone lower as the pilot took the cut and I watched the outline of the night-fighter arrest on the first or second wire. As the first Hellcat taxied forward to just beneath my perch, the second one set down in a textbook landing.

"For all the hours of training and instruction I had received, I had not really believed it possible before seeing it with my own eyes.

"We gathered around the two Navy veterans and asked a set of rapid-fire, nervous final questions. Then someone with a bullhorn blared: 'Fly One, Marines, man your airplanes.'

"It had been decided that only one of us would launch and recover at a time, until everyone had completed one landing. I was the senior one in the group, so I would launch first. I did my best to hide my terror, and I think the only reason I

was not found out was that everyone else was too deeply involved in masking their own fright to notice mine. My flight gear was still in the cockpit of my Grumman, so I climbed aboard and quickly snapped snaps and pulled straps. I recalled my first real moment of truth in an airplane, my qualifying solo at the elimination base at Long Beach, California in 1940. Wow! I had come so far. I had been eager to fly then and had remained eager as I passed every milestone to this moment, when I found myself unforgivably apprehensive for the first time in my flying career. The prospect of first combat had not come as close to terrorizing me as this flight. I turned up the engine and allowed myself to be guided to the catapult. Before giving the catapult officer the 'ready signal', I nervously checked and rechecked my harness, pulling the straps again for good measure. Then I ran through my pre-flight checklist; screw the men in the island who were moaning about how long I was taking: canopy back and locked, engine at full r.p.m., prop at full low pitch, flaps all the way down for maximum lift, right foot hovering above right rudder pedal ready to overcome the left torque of the spinning prop, stick held loosely in my right hand, throttle grasped loosely in my left hand, head resting against the headrest to take up the shock of the catapult.

"I looked to my left and saluted. Ready! In response a dimly-perceived deckhand standing over the catapult crew's catwalk whirled a flashlight. Go! I turned my eyes front, loosened my grip on the stick, set my jaw and leaned back into my seat.

"WHACK. My conscious mind was aeons behind my senses, as it had been on all previous catapult launches. I had a very busy couple of seconds as I kicked the right rudder pedal and yanked the stick into the pit of my stomach. I had no time to dwell on how dark it was out there.

"My equilibrium returned. The Hellcat was climbing away to the left. I got the wheels and

If he were any more stupid, he'd have to be watered twice a week.

It's hard to believe that he beat out a million other sperm.

flaps up in one motion. I had that familiar short sinking sensation as the flaps went up and the Hellcat dropped slightly. Then my mind kicked in: 'needle — ball — airspeed, you dumb cluck.'

"A destroyer passed beneath my low left wing. I had just enough time to notice two blinking navigation lights before the inky black of the perfectly dark night enfolded me.

"All my training and experience saw me through a climb to 3000 feet. While my mind reeled off a thousand facts about my flying, my voice talked to the ship in calm tones, reporting on routine matters the air officer would want to hear about. I was neither here nor there.

"I was cleared to land, which was of both relief and concern. I wanted to get down but first I had to find the carrier.

"I coaxed the Hellcat into one full circuit of the area in which I was pretty sure I had left the *Tripoli* and her escorts. I saw the two navigation lights on the forward destroyer I had overflown after launch. Then I saw two more which had to be the plane guard destroyer deployed a few thousand yards directly astern of the carrier, to pick up downed 'zoomies', as we carrier pilots were known.

"Now I knew exactly where the flight deck lay. I also knew that, in the event of extreme danger, the carrier flight deck lights would be flicked on to help me find a safe roost. But that would mean failure, and there were too many people watching to let that happen.

"After reassuring myself that I was flying on a heading opposite that of the ship, I flew down the carrier's port side and approached the plane guard from ahead, keeping it just off my port wing. I could not help ruminating about how useless a night search for a bilged aviator must be.

"I flicked on my radio altimeter, a brand new instrument that had been installed in my cockpit just before we left San Diego. I had set it for 150 feet. If I flew above that altitude, I'd get a white signal light. If I flew below 150 feet, I'd get a red danger light, and if I was flying right at 150 feet, I'd get a comforting green light. The light was green when I turned on the altimeter.

"I flew upwind the length of the tiny destroyer and sighted her deck lights, which could only be seen from the air. This was the only concession to a pilot's natural tendency to become disoriented across even the briefest interval of night space.

"I had been timing my flight ever since passing the carrier and spotting the plane guard's lights. At what I judged to be the best moment, I turned 90° port, dropped my wheels and flaps, enriched the fuel mixture, partly opened the cowl flaps, put the prop in low pitch, and turned another 90° to arrive at a downwind position dead astern of the carrier.

"My night vision was, by then, as good as it would become. I had been training myself to find dim objects with my peripheral vision, which was the preferred method. Thus, I was able to pick out the dim shape of the totally darkened carrier as I floated up the wake.

"I was committed to the approach. All my attention was focused on sighting the LSO's luminous paddles. I momentarily panicked and said, or thought I might have said, 'Where the hell are you?'

"First, I sensed the coloured paddles; then I knew I saw them. The LSO's arms were both out straight. Roger! My ragged confidence was restored, though I was a good deal less cocky. I checked my airspeed, which was down to the required 90 knots. Before I knew it, I saw the cut. The tailhook caught a wire and I was stopped.

"I taxied past the barrier, came to a rest beside the island and cut my engine. As had been the case after my first combat mission, my flight suit was reeking of sweat.

"During the rest of the night, my eight fellow fledglings each made one night landing. My subordinates accounted for more than a few wave-offs, but that was partly my doing. I had asked the LSO to be particularly unforgiving of minor gaffes.

On 23 August 1943 we departed Pearl Harbor with the USS *Yorktown* (CV-10) for our first Pacific area combat operation, a strike against Marcus Island only 700 miles from Tokyo. On 1 September we were launched pre-dawn from about 150 miles north of Marcus. A pre-dawn from a completely blacked-out carrier is an incredible experience. You take off into the pitch black night with no running lights on any of the planes or other ships in the task force. You climb up toward the rendezvous point and finally see the exhaust flames from a slightly darker blob ahead. You join up, not knowing who it is since you cannot see the numbers on the other plane. When it finally gets light enough to see numbers, you shuffle around to your proper place. That morning, after we had finally gotten all the strike force rendezvoused and on the way into Marcus, some idiot towards the rear of the group accidentally (I hope) fired all six .50 caliber machine guns, sending an arc of huge red tracers over the strike force. In the dim dawn light a .50 caliber tracer looks about the size of a basketball, and more than a few planes took rather violent evasive action.
– Hamilton McWhorter, former US Navy fighter pilot

Above: Many things have changed over the years in the Navy. The LSO with paddles of the World War II era has given way to one who communicates and controls the flying performance of landing pilots electronically. Pilots guide themselves to a landing by monitoring the position of an illuminated "meatball" display near the port side of the flight deck, while the landing signal officer watches, signals, and evaluates.

We all knew how important it was to get this exercise 100 per cent perfect."

"It has been said that those who stand and wait also serve. I didn't fly combat because there was no war. Like countless others, I was there if they needed me. I had to be content with peacetime operations, and I was. It excited me to fly the meatball or make paddles passes, to launch from a deck run over the bow or, once in a while, get a cat shot, and fly the downwind leg, judge the 180° turn in a stiff tailwind and come aboard smartly in a prop-tailwheel type aircraft, like my heroes in F6Fs and F4Us did in World War II. I

was a kid, hoping to follow in their footsteps if there had been a war. Few are eager for a war, yet every kid joining the fleet thinks he is bullet-proof and that, if war comes, he'll just follow his leader and do whatever he does.

"Everyone remembers his solo flight, his first emergency, his first girl, first car and so on. I ate up six months of peacetime carrier flying as though nothing else in the world mattered. In some ways, nothing has ever topped that, other than the love of a wife and the births of healthy children. Thank God there wasn't a war on when I served, but when you're in uniform, you serve. I am damned proud of what I did."
– Paul Ludwig, former US Navy attack pilot

"I first met AMM3/c Peter Gaido as I was preparing to make my first carrier landing. He asked, 'Can I go with you?' I replied, 'This is my first carrier landing and I am supposed to have only sand bags.' He said, 'You got wings, ain't 'cha?' and replaced the sand bags with his stout frame. With that supreme vote of confidence, I made half a dozen perfect landings.

"Peter later displayed his character as he observed a Japanese bombing plane attempting to crash into the *Enterprise*. He jumped into an empty SBD, fired the machine gun at the approaching plane and continued firing at it as it sheared off the tail of his SBD. He continued firing as the Japanese plane moved in the opposite direction until it hit the ocean. He tried to remain anonymous in this action, but Vice-Admiral Halsey finally found him and promoted him on the spot to AMM1/c.

"He flew with Ensign O'Flaherty during the Battle of Midway. They dropped their bombs on the Japanese carriers but later had to make a crash-landing. They were picked up from their rubber boat, rescued by the Japanese, interrogated, murdered, their bodies thrown into the sea.

"Peter Gaido was the bravest man I ever met."
– Jack "Dusty" Kleiss, former US Navy pilot

LSO

THE LANDING SIGNAL OFFICER, or LSO (Batsman in the Royal Navy), is the person who, for much of the time that aircraft carriers have been in use, has been responsible for guiding pilots onto the flight deck to a safe landing or "trap". Each squadron on today's US Navy carriers has its own landing signal officer who assists his fellow aviators down the approach, confining his communication to a minimum if the pilot is doing it right. If he isn't, the LSO will signal corrections to the pilot, and/or talk him down with "candy calls" until he makes the necessary adjustments. If he fails to properly correct, the LSO will give him a wave-off, a signal which must be obeyed, sending him around to rejoin the pattern and try again.

In the US fleet today, the carrier air traffic control center (CATCC) guides returning aircraft to the carrier control area around the ship. When planes return from missions and adverse weather obviates a safe visual approach to the ship, the CATCC controls their arrival. It clears each aircraft for approach at one-minute intervals. The landing signal officer then comes into play, helping the pilot in his or her final approach to a landing. A seasoned naval aviator and carrier pilot, the LSO has been thoroughly evaluated and well-trained at Landing Signal Officer school and on board ship. The LSO is an expert naval aviator who must bring sensitivity, the wisdom of experience and exceptionally good judgement to the job. He or she performs this vital function using state-of-the-art equipment from a platform area adjacent to the landing area, on the aft port side of the carrier. In this prime viewing spot the LSO, assistants, and LSOs in training, monitor wind and weather conditions, characteristics of the operating aircraft and the motion of the deck. They must also take into account the experience level of each approaching pilot in turn.

In ordinary visual flying conditions, carrier aircraft return from their missions and may be positioned in a "marshal stack". When this happens, flight leaders take their interval on flights at lower altitude levels in the stack. The actual landing procedure begins as either two- or four-plane formations enter the break for landing, astern of the carrier, on the same heading and to the starboard side at an 800 foot altitude. As the flight leader reaches a projected point ahead of the ship, he will break left and align his airplane on a downwind leg while descending to 600 feet and completing his landing checklist. On the final approach leg he will rely on the carrier's automatic, gyrostabilized Fresnel lens optical landing system — an arrangement of lenses and lights positioned off the port edge of the angled deck. If the carrier is rolling and pitching beyond the limits of the gyrostabilization capability, or if the Fresnel system should fail, a manual optical–visual landing aid system (MOVLAS) can be quickly set up for use. When the weather cooperates and conditions for aircraft recovery are reasonably good, the LSO operates "zip-lip": without radio communication.

On his final approach to the deck, the pilot is flying at between 120 and 150 knots, depending on the type of aircraft. To use the Fresnel system, he must first locate the array of lights and focus on the amber light or "meatball" in the centre of the mirrored lenses. If he has properly aligned his aircraft on the glide slope, he will see the "ball" aligned with a horizontal line of green reference lights on either side of the centre lens. If his aircraft is too high (above the glide slope), the ball will appear in one of the lenses above the centre lens; if the aircraft is too low, the ball will show on one of the lower lenses. An optimal carrier landing requires that the pilot visually keep the ball centred all the way down the glide slope to touchdown on the deck, and then engage the "three wire", the third of four heavy cables which are stretched across the aft area of the flight deck from just ahead of the ramp or rounded aft end of the deck. It is the pilot's responsibility to "fly the ball". The LSO can and does help him down with light signals and/or voice instructions.

On 10 September 1952, while embarked in HMS *Glory* in the Mediterranean, I was briefed to carry out a take-off using Rocket Assisted Take-off Gear (RATOG). The brief required the pilot to carry out a full-power take-off firing the RATOG at a specific point on the take-off run. The brief then said that if the RATOG did not fire, there would be room to throttle back and brake to a stop before reaching the bow. The brief should have said, there would be nearly room . . .

When I pressed the button to fire the RATOG, it didn't. I throttled back and tried to brake to a stop. I did not stop and instead went over the bows at a walking pace.

Eyes shut, eyes open, just as quickly, to find the cockpit full of water and bubbles just the way it was in the USN "Dilbert continued on page 175

Left: Sometimes referred to as "paddles", a landing signal officer from World War II.

Below: The pilot of this Grumman E-2 Hawkeye is guided to a safe trap while being observed by the duty LSO of this group on a US Navy supercarrier in the 1990s.

For all that has been said of the love that certain natures (on shore) have professed to feel for it, for all the celebrations it has been the object of in prose and song, the sea has never been friendly to man. At most it has been the accomplice of human restlessness.
– from *The Mirror of the Sea* by Joseph Conrad

When a pilot has flown his plane to a point roughly three-quarters of a mile from the carrier in the final approach to the deck, the ship's air traffic control centre delegates control of the aircraft to the LSO. The LSO and his half-dozen or so assistants, many with binoculars trained on the incoming plane, check first to be sure that its landing gear and flaps are in the proper down position, that the plane is lined up on the centre-line of the flight deck, that its wings are level, and that it appears to be descending at the correct rate to catch one of the four arresting wires when it slams onto the deck.

Bringing a hot and heavy beast like an F/A-18 Hornet aboard a carrier is a precise and demanding task, with little margin for error. The pilot must establish his plane's position and attitude on the glide slope and then fly it down the slope at an exact three-and-a-half degree angle. If, when he

arrives over the ramp, he has flown the slope perfectly, his tailhook will cross the ramp 14 feet above it. If he is more than a few feet too high, his hook will miss the arresting wires, or bounce over them, for what is known as a "bolter". Procedure requires Navy pilots to shove the throttle (or throttles in multi-engine aircraft) to full-power the instant the airplane contacts the flight deck. When the plane's tailhook catches an arresting wire, bringing it to an abrupt halt, the pilot immediately retards the throttle, the plane is allowed to roll back a few feet so the tailhook is disengaged from the wire, the hook is raised and the aircraft is guided away from the active landing area. The arresting gear is then reset for the next approaching plane. Should a pilot experience a bolter and his tailhook fail to catch an arresting wire, his engine is at full-power so he can get airborne and go around for another landing

attempt. With every approach, the LSO and his back-up each hold an up-raised "pickle" switch which either of them can use to activate flashing red lights on the Fresnel array, signalling a wave-off to the pilot. The wave-off is not optional.

LSOs grade every landing approach to a carrier. They use a Trend Analysis form to chart the ongoing record of every pilot's performance. When all landings in a flight operation have been completed, members of the LSO team visit the pilots who have just landed, to discuss their grades for the task. Few things in the life of a carrier aviator are more important than these grades. If he cannot consistently and safely bring his airplane aboard the ship, the pilot is all but worthless to the Navy and will be sent to the beach. The grade given by an LSO for a landing approach is nearly always final and not subject to appeal. The grades given range from OK: a good approach with no problems; to FAIR: a performance with slight deviation from the correct approach; to NO GRADE: unacceptable deviation from the correct approach, to WAVE-OFF: the approach was too far from correct, was unsafe and had to be aborted, to BOLTER: try again. Competition among pilots for good grades is naturally high. When a pilot has done everything well and made a good approach, his tailhook should catch either the number two or number three wire. If the approach was a little high or fast, he may catch the number four wire; if a bit low or slow he may get the number one. Both numbers one and four are less safe than two and three — three being the most desirable.

One of the most dangerous things a human being can do is a "night trap": landing on a carrier at night. Nothing a pilot must do requires greater skill, is more demanding and genuinely frightening. Poor performance in carrier night landings is the biggest cause of naval aviators losing their wings, and some pilots say that even after hundreds of night carrier landings, they never get easier. Some feel that the more night landings they do, the more nervous and uneasy they are about them. "Doing a night trap concentrates the mind wonderfully", comments one F-18 pilot. It is the lack of visual cues that creates the problem for pilots on night traps. Landing at night on an airfield offers a whole range of vital cues with which a pilot can judge how he is doing in his approach. Night-time at sea is black on black. Often there is no horizon, making the experience even more disorienting. Setting up a proper approach can be a genuine waking nightmare. However, many people believe that a little fear (in combination with intelligence and a high degree of capability) in such a demanding situation, can be a good thing. It keeps you sharp, focused.

When carrier-based planes have to come to the boat in foul weather, there is a wide spectrum of electronic aids ready to assist them on modern US carriers, from the instrument landing system (ILS), to the tactical air navigation system (TACAN), to the carrier controlled approach (CCA), to the automatic carrier landing system (ACLS). The latter is capable of bringing a plane to touchdown when the pilot has no visual contact with the ship, much less the flight deck. In ACLS, a precise guidance radar on the ship locks on to the automatic pilot in the plane when it is eight miles out from the carrier. Computers on the carrier and in the plane feed position updates to each other and the ACLS sends signals to the plane's autopilot which establishes the approach. With this method, the autopilot can fly the plane to a safe landing on the carrier without the pilot having to touch the stick.

On 1 May 1945 several US Marine F4U Corsairs were operating from Kadena airfield on the island of Okinawa in support of troops there. The carrier USS *Yorktown* received a "May Day" distress call from three patrolling Corsairs that had been blown off course and out to sea by unusual 100 m.p.h. winds. The fighter pilots were dangerously low on fuel when they called for help. When the carrier

Dunker" ditching trainer eight years before. Unstrapping my seat harness and parachute harness I tried to stand up and clear the cockpit. As I stood up two spigots in the cockpit canopy, designed to slot into holes on the windscreen, caught the back of my Mae West and I was unable to clear the cockpit because of the water pressure caused by the rate of sink of the aircraft. I had to get back into the cockpit and turn round so that I could get out by covering the spigots with my hands and force myself clear. Once clear I saw the aircraft turn onto its back as it continued to sink on down, but there was no sign of the ship or its propellers which frightened me somewhat. My first idea was to stay down until the ship passed over me, but then I realised that if I stayed any longer where I was I would probably drown.

I was then confused as to which way up was, but that was cleared by inflating my Mae West. I tried to swim up and away from where I thought the side of the ship was. I surfaced about ten yards clear of the ship's side abeam of the starboard propeller. I immediately took a deep breath and choked as I still had my oxygen mask on and the oxygen tube was under water.

After unclipping my mask I noticed that my dinghy was still clipped to my backside. It was a work of moments to unclip the dinghy, inflate it and clamber in. After that all I had to do was sit there in relative comfort and wait for the seaboat from the plane-guard destroyer to come and rescue me.

– Alan J. Leahy, former Royal Navy fighter pilot

A safety net, perhaps ten feet wide, was placed a few feet below and abaft of the LSO. From "Vultures Row" I watched LSO Bert Harden signal a plane to take a wave-off because the aircraft was too slow. The airplane responded slowly, and headed directly for Bert, who dived for the safety net. He jumped so vigorously that he completely missed the net and made a perfect swan dive into the sea 65 feet below. The destroyer behind *Enterprise* picked him up unharmed.
– Jack "Dusty" Kleiss, former US Navy pilot

answered their call, the Marines reported that none of them had ever landed aboard a carrier before. "Do you have tailhooks?" asked the *Yorktown* officer. "We'll see . . . Yes", replied one of the Marine pilots. The carrier's Air Boss then obtained permission from the ship's captain to let the Marines try to land aboard. "Crash Two", the secondary Landing Signal Officer on *Yorktown,* got on the LSO platform radio and talked the Marines into the landing pattern and up the groove, while the primary LSO, "Crash One", signalled them with his paddles. All three pilots landed aboard safely and as one of them climbed out of his cockpit, he remarked to one of the airplane handlers, "What was that man doing waving those paddles back there?" "Brother, he's the Landing Signal Officer and he was giving you a wave-off!"

Skyraider pilot Paul Ludwig believes that the problems he encountered during his Navy tour in the 1950s were "small potatoes compared to the great and wonderful privilege of flying high-performance military aircraft and getting paid to do it. Forty years after my tour on the *Hornet,* I still remember the thrill of being a carrier pilot. I tend to downgrade the incidents. It was the thrill of a lifetime to have completed a tour on a carrier. Nothing was more exciting to me than making a paddles pass in coming aboard the carrier and catching a three-wire. The *Hornet* had the meatball — the mirror system — but occasionally, we got aboard by paddles. The mirror system is only a gadget, and gadgets break. The thrill of having an LSO waving signals to me, and following his signals, was something I had seen in *Victory At Sea*, *The Fighting Lady* and *The Bridges At Toko-Ri*. I grew up in World War II, longing to make a landing on a carrier with the help of an LSO. All I ever wanted to do was what those in combat had done. What I saw in those movies were men who had survived the battles of a particular day, and were making paddles passes, getting a wire, taxiing

forward, shutting down, getting out and walking across the deck looking great. It was those movie heroes who made me seek the fulfillment of my dream."

In his extraordinary book, *Daybreak for Our Carrier*, Max Miller described with great clarity the role of the LSO on one of the US Navy's fast carriers in World War II.

"The job aboard a carrier which appears to be the most fun (from a distance and in photographs) is that of the landing signal officer. He is the one so often pictured standing aft by the fantail, colored paddles in each hand, and waving them as if the air were rent with hornets.

"His base of operations is a small grilled platform which swings off the flight deck, and his backdrop is a screen of canvas. He doesn't stand in front of this backdrop when he is signalling, nor for that matter does he stand on the grilled platform. But they are his base of operations nevertheless, and the screen of canvas also serves as a windbreak.

"There is a slight trace of the bullfighter in a landing signal officer. He not only has to know how to lure 'em on with his colors, but he also has to know how to jump should things get too hot.

"He himself is a carrier-trained pilot. Why he was selected from regular daily flying duty to be a landing signal officer is something he most likely will avoid answering outright. It may be that he really doesn't know. Or it may be that the powers that be told him they saw in him exactly all the stuff that a landing signal officer should have. And other than that, he had no say in the matter. This is the most likely.

"Planes cannot simply race in and land aboard a carrier without guiding help. They could try it, of course, and supposedly there would be some which would succeed. But on approaching the flight deck for a landing, the pilot encounters a most definite blind spot. He cannot see his own wheels at any time, of course, nor can he see the immediate spot directly beneath him and directly in front of him where his wheels should first touch. The landing signal officer has to be the other pair of eyes.

"That is one reason.

"Another reason is that, with the planes of three squadrons circling the carrier ever lower and lower for their landings, somebody has to be at a conspicuous spot to direct the timing between planes, and to see to it that the flight deck doesn't become one beautiful mess of tangled-up propellers.

"It must be remembered that a flight deck, though massive both in appearance and in actuality, is nevertheless limited in space where the planes must make their landings. This space aft with its arresting gear and barriers is less than half the deck's length. To over-shoot this space, and to try to make a landing anyway, would mean to crash into the planes already aboard and which have been brought forward beyond the barriers as fast as they can be brought.

"The planes, on their approach 'in the groove,' come in at a speed of about seventy or eighty knots. At least this is the speed which is figured on if all is going well. Coupled with this, the ideal head-on wind for landing or launching planes is between thirty and forty knots. This does not mean that the wind itself has to be that strong literally. But the head-on speed of the carrier into the wind is making up for some of it.

"So, when a landing signal officer is standing out there with his colored paddles, and a plane is approaching him for a landing, he has to keep a lot of things in mind besides just the plane itself. He has to keep the wind in mind, as mentioned, and he also has to keep in mind the condition of the flight deck on his side of the barriers.

"But with his eyes concentrated on the incoming plane, and with the pilot of the plane concentrating in turn on the signal flags, it could all become quite a jumbled-up affair if the signal officer took

time off to gaze at the condition of the deck behind him. He does, then, have his assistants, and one of them, an enlisted man, is the 'talker.'

"The 'talker', with earphones and a mouthpiece, squats just over the edge of the flight deck so that his eyes are level with it, and he keeps watching what is occurring to the plane which has just landed a few seconds previously.

"If there is difficulty in getting the plane released from the landing-gear, or if there is difficulty in getting the plane taxied up forward beyond the barriers, the 'talker's' conversation to the signal officer is all one-sided. It consists of the repetition: 'Foul–foul–foul–foul–foul–' and then possibly a 'clear.'

"Yet it is at such points as this that a signal officer has to make one of his many split-second decisions. He is as anxious as anybody in the ship to get all the planes aboard as soon as possible. He doesn't want to give the next incoming pilot the good old 'wave-off' any more than the incoming pilot wants to receive it.

"But if, at that critical moment of timing between speed and distance, the 'talker' is still saying 'foul,' then aloft go the signal officer's flags in criss-cross waving fashion. The pilot, in that same split second too, must push on more power to bank-turn over the deck and zoom away.

"Though pilots are obliged to take a 'wave-off' whether they like it or not, unless something devilish is the matter with their plane, they are not obliged to land even after the signal officer signals the 'cut' to do so.

"The signal 'cut' means, of course, to cut the motor and let the wheels touch. The signal is indicated by a quick cross whip of the flags down low. It's then up to the pilot to do the rest. The signal officer is through with him, and now looks for the next incoming plane. The 'cut' is the gesture finale with each plane, and with it the plane whirs on past the signal officer onto the deck for better or worse.

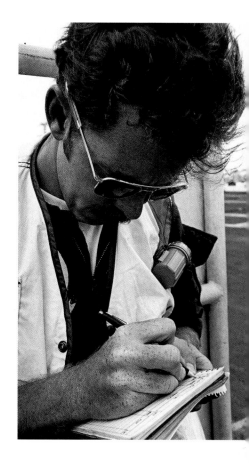

Above and left: LSOs at sea off San Diego aboard the US carrier *Coral Sea* (CV-43) during routine carrier landing qualifications for pilots in the summer of 1974.

Top: An F/A-18F Hornet traps on the USS *John C. Stennis* (CVN-74) in January 1997. Bottom: A Fresnel Optical Landing System light arrangement on a US supercarrier.

"There's a phrase which is used at times when a pilot, after getting the signal 'cut', makes a bad landing through what appears to be his own fault. So dependent has he been on the flags of the signal officer during the past few seconds that now, when suddenly on his own, the phrase is: 'He stopped flying.'

"Another job of the signal officer, as if his hands weren't filled enough, is to grade each landing much as a schoolteacher would do. This goes for the approaches as well. The moment the plane is aboard, and before the signal officer has time to forget the mental picture of the approach and of the landing, he quickly rattles off an abbreviated code of his own describing his opinion. An assistant scribbles the letters into a notebook after the pilot's name or after the number of the pilot's plane.

"Afterwards, and usually when the pilots are in their ready rooms getting out of their togs, the signal officer from his notebook will tell them how they did, or what they didn't do, or what they should have done. All of which can be of help for the next time.

"But the art of signalling planes aboard, and bringing them aboard rapidly with the maximum of safety, is such a complicated art that even when we watch from the island we do not catch the full picture. Or at least we do not catch it in the true perspective. The only place to catch the true perspective actually is right down there in the net next to the signal rack itself.

"From the high island the perspective is such that what appears to be the making of a crack-up for sure may turn out in the end to be a very nice landing indeed.

"From up in the island too, as in the grandstand of any football game, we are tempted each time to be the quarterback. Or in this case, the signal officer. Why doesn't he give the 'wave-off?' Or: Why doesn't he give the 'cut?' Or: Why did he give the 'wave-off?' It looked all right to me.

"Yes, we cannot resist being unofficial signal

officers, none of us. All of which adds to the life aboard a carrier, too.

"There's the story which goes how some new captain on one of the smaller carriers had much the same idea about quarterbacking from high up on his distant bridge.

"Through the loud-speaker the new captain harassed the signal officer so much during the landings, and began yelling to him so much what to do each time, that the signal officer suddenly had to decide between smashing up the planes or his own Navy career.

"He was so blindly furious about all the dictation during the height of a landing that, using an artist's prerogative (and he certainly was an artist), he tossed his signal paddles onto the deck and went below.

"He aimed for the empty wardroom, and stayed there drinking coffee, trying to drown with it what he was thinking. Meanwhile, the remainder of the planes continued circling and circling the ship waiting for the signals, the pilots wondering what the hell.

"Perhaps the end of the story might have been different if competent landing signal officers were something which could just be picked up for the asking. But months of training, and even years of training, have gone into what they do which may appear so easy. And in addition to their training they also have to have that little 'something else' besides to be classed in the limited group of the truly top-notchers. Their fame, though not known to the public, is certainly known from carrier to carrier in this Pacific.

"A top-notcher, though he may classify himself as 'just another of those plane-bouncers' is really a gifted personage. It is taken for granted that he can signal the planes on in 'along the groove,' that with his paddles he can talk with the pilots continually, that he can tell them they are too low or too high or at too much of an angle or too fast or too slow. It is taken for granted, also, that he is

responsible for making sure their wheels and their flaps are down before they come in. All this part of his ordinary work is understood.

"But a true top-notcher is one who can go beyond any of this. He is the one who knows the personal characteristics of the individual flyers aboard. Some of these flyers, he realizes, are better at one type of approach than another, and some are just naturally so good at carrier-landings that he need not worry too much about them, but can concentrate on the others instead.

"If there is to be uniformity in the landings, naturally, it is well for him to see to it that all the flyers behave more or less the same way. But in those cases of emergency, in those cases where planes have to be brought aboard regardless, and brought aboard fast — these are the moments when the artistry of a signal officer really shows, and really pays dividends.

"He knows in a second what allowances to make for one flyer, and what not to allow for another. Some can get by, and skilfully, with something which might cause others to hit the deck too hard. He will recognize the pilot by the number of his plane, or he may recognize the pilot himself as he circles by. And when a plane has signalled the instant need of an emergency landing regardless of anybody or anything, but preferably a landing on deck, all this depends too on the signal officer's ability. Or when they come in with their big bomb loads still stuck to the plane and unreleasable. Or when they come in with their landing-gear shot away. Or when they come in wounded and barely able to make it. These are the moments when a true top-notcher, working his most delicate best with the pilot, is surely an artist supreme.

"Anyhow, to return to that story which was started some while back, it was not long before that new captain on the little carrier began imitating the pilots aloft by also wondering what the hell.

"There was no other landing signal officer aboard, so now you know how the story ends. He was

The following incident occurred on the USS *Constellation* (CVA-64): An F/A-18 pilot, call sign "Oyster" was launched from Cat 1 and fodded (foreign object damage) both of his engines. Most people witnessing the launch thought that the pilot had ejected, but soon realized that he was still in the plane. Somehow, he managed to get the Hornet level at about 80 feet and then gradually milk it up to 150 feet. It was 8.45 p.m. and there were scattered clouds with a ceiling between 1000 and 1500 feet. It was quite dark.

One of the Hornet's engines had failed completely, while the other was having massive compressor stalls which were obvious to personnel on the carrier, even at a distance of 6 miles. The pilot was maintaining his minimal altitude through the use of full afterburner on the one remaining engine, with his landing gear up. He jettisoned his weapons stores and dumped his fuel down to a level of 4.0 in order to maintain level flight. He was unable to climb to 3000 feet in order to do a wave-off/ approach capability check. However, the flight deck was ready to recover him, and the ship decided to let him attempt a normal pass. He began to line up for it, and immediately decided that he could not make it successfully. He came up along the starboard side of the ship, and again, it appeared that he was going to eject. An incredible amount of debris was coming out of the right engine of the Hornet, and by this time the pilot had run his remaining fuel down to near zero. It *continued on page 180* seemed that he would be able to adequately slow his rate of descent once he began his approach to the carrier, but he clearly would have no bolter capability, and it was decided to erect the net-straps barricade to stop his aircraft if necessary. The barricade was rigged quickly, and the LSO platform was cleared except for the landing signal officer and his back-up.

The F/A-18 was high in the approach and, by the time it had begun to correct down, the pilot was getting too far out of the parameters and the LSOs had to "pickle" him (signal a redlight wave-off). The Hornet staggered past the ship, its one engine making a sickening whine and pop as it emitted a salvo of flare-like matter from the tail cone. Clearing the top of the barricade by less than 15 feet, the plane was now down to .8 on fuel. The pilot was able to climb to just 600 feet for his last approach and the LSO talked to him continuously as he intercepted the glide path at one mile out from the ship. The LSO told the pilot to go ahead and sacrifice a few knots of air speed if necessary, to keep the airplane on the glide slope, as the ship had plenty of natural wind over the deck.

The pilot was having trouble keeping the Hornet aligned in the approach, and drifted left a bit, and then was a little low, but, as the plane neared the ramp, the LSO felt that the pilot was going to clear the round down, and gave him the "Cut" to land. The plane slammed onto the deck about 20 feet beyond the round down, and went on to

summoned from his coffee back up to the flight deck. Nobody coached him over the loud-speaker after that. Nor, according to our version of the story as told to us, did anybody mention court-martial.

"For a landing signal officer is at his best when, along with knowing the rest of his trade, he has the absolute confidence of his flyers — and this one had it. They stuck by him the same as he in turn had stuck by them by not heeding distant advice. If the confidence in a landing signal officer ever has cause to become the least bit wobbly, a pilot consciously or subconsciously may hesitate about the signals at a critically wrong time.

"So all in all a landing signal officer, in his yellow sweater and yellow cloth helmet, may look gaudy out there next to the fantail, and he may look funny waving those colored paddles around his head. But above him there may be as many as sixty pilots and their gunners who would like very much to be able to eat that night. He wants to see to it that they are able."

Robert Croman was an SBD Dauntless dive-bomber pilot in the South Pacific during World War II. In March 1944 he was ordered back to the US to attend Landing Signal Officer School. After graduating, Croman was assigned to Carrier Air Group 19, which was to be the first US Navy unit to fly the Grumman F8F Bearcat.

A heavy fog had suddenly materialized at the US Navy airfield near Santa Rosa, California and visibility there was less than half-a-mile. It was 18 July 1945 and Bob Croman was the LSO for CAG 19, which was practising field carrier landings for the first time with their new Bearcats. The pilots of 19 were discovering that the new plane was both extremely powerful and utterly unforgiving. Croman was in position at the end of the runway. With him were an assistant LSO, an ambulance and its crew.

The hot little Bearcat, with its big Pratt & Whitney

engine and huge prop, developed a lot of torque and a pilot had to keep 'ahead' of the machine or it would get away from him.

The first three planes did well and Croman gave them each a "cut" signal to land. The next pilot was clearly losing control of his plane in the approach, coming in too slowly. The LSO gave him the "slow" signal, but the Bearcat continued lower and slower in the approach. At that point, Croman tried desperately to get the errant pilot back on track as the fighter roared towards him. The pilot added power, too late. Torque pulled the plane to the left, towards where the LSO stood. It hit the runway hard, collapsing the left wing and landing gear, and twisting the big propeller.

Had this happened aboard a carrier, Bob Croman might have been able to throw himself into the safety net adjacent to the LSO platform. On this airfield, there was no escape for him. The Bearcat slid, screeching towards him; he dove at the area beneath the plane's right wing and felt a heavy thud. He was lying on his stomach, dazed and unable to move. To Croman, it seemed that he had lain there a very long time. The wrecked plane had continued down the runway after hitting him and the pilot got out of it without so much as a scratch.

The LSO had been flung like a rag doll by the impact with the Bearcat. He lay contorted, with his right leg mangled. The leg was bleeding profusely and he was obviously in need of urgent treatment. People at the scene of the accident rushed to his aid. One person applied a tourniquet and removed Bob's jacket. A medic gave the LSO a shot of morphine. Oddly, up to then Croman had felt no significant pain. Suddenly, the pain arrived and was unbearable. Still, he says: "I knew I was not dead. You can't hurt that bad and be dead."

The doctors at the Santa Rosa base dispensary could do little to help apart from stabilizing the LSO. His leg was nearly severed six inches from the ankle, attached only by bits of flesh.

Croman was flown to the Oakland Naval Hospital and rushed to an operating room. He lay there awaiting help from a doctor when the doors opened and in came a Dr McRae, the same Dr McRae that Croman had met and become friendly with fifteen months earlier out in the South Pacific. The doctor took charge of the situation and reattached the LSO's leg in a surgical procedure lasting more than eight hours.

Bob Croman underwent sixteen additional operations on his leg during the five years after the tragic accident. Both his Navy and flying careers were ended. But, despite the fact that his leg has atrophied and he cannot move his ankle, he is extremely grateful to the Navy for the treatment he received following the accident, and especially thankful to Dr McRae who somehow arrived at that critical moment.

"The threat is as much the boat as the enemy. Carrier aviation is a very unforgiving occupation.

Lives are lost and careers destroyed due to the specialized skill required to land on the boat. There are many excellent pilots who can fly a fighter with incredible skill but can't land on the boat. And the opposite. One study of experienced carrier pilots who had been hooked up to sensors, showed that their heart rates and respiration were actually higher when landing on the ship than in combat. Daytime carrier landings are generally not too difficult if the seas are calm and the weather is tolerable. Night carrier landings are always challenging. Operating on the ship at night is not a soothing experience. It requires complete and total concentration along with a good measure of confidence. Night take-offs and landings during poor weather with a pitching deck, and blue water ops with no divert are the ultimate tests of aviation skill. I would question the credibility and sanity of anyone who claimed to be relaxed during night traps.

"Your first carrier landing is an experience you never forget. In my case, the training carrier at the time was the USS *Lexington*, an old World War II

engage the barricade on centreline. It was an emotional moment on the LSO platform and around the carrier, with great cheers erupting on the flight deck as Oyster popped the Hornet canopy and emerged safely. His airmanship, guts and determination in saving his airplane had been heroic.

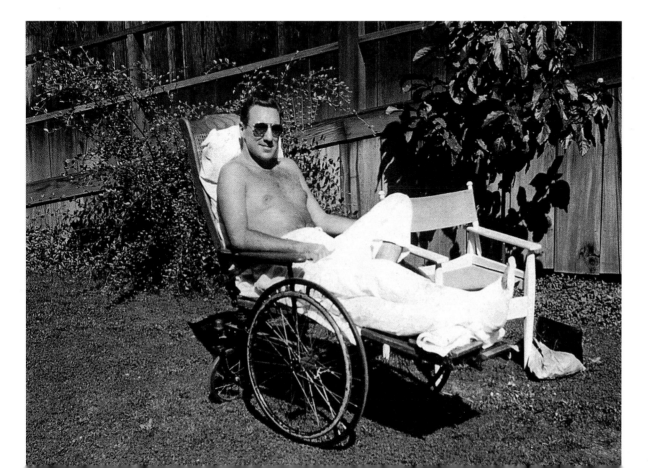

Left: Robert Croman, a World War II LSO, after surgery to reattach his right leg which was nearly severed in a landing accident.

Below: A plane director helping to clear the deck after a landing. Bottom: Not all landings are happy. Here the airframe of an AD Skyraider of VA-45 burns in a pool of fire-retardant foam following a carrier landing accident in the 1950s. Below right: A US Navy landing signal officer at his job during Operation Desert Storm, the Persian Gulf, in early 1991.

era small-deck carrier that was based at Pensacola, Florida. We had a very early brief around 0530. The night before we had all tried to get to bed early in order to be rested for the day. My roommate was asleep in about fifteen minutes and began a loud and continuous snore. I couldn't relax and got two hours sleep. When we reported to the brief, the flights were delayed for hours, but we had to 'stand by'. It was going to be a long day. Your first carrier landing is solo. An instructor pilot takes out three students on his wing and drops them off in the 'break'. The break is a manoeuvre that starts parallel to the ship on the starboard side at 800 feet. Once past the bow, a hard level turn and descent to 600 feet for the downwind. At this point you are slowing the aircraft and configuring for landing, speed brakes, gear, flaps and hook. Spacing is critical, not only on the ship, but relative to the aircraft ahead of you. The Air Boss wants 45-second intervals for landing. The speed in the break is usually around 350 knots. You are on your own from there until you finish your qual. Your first two passes are touch-and-gos. The intent is to put young pilots at ease, knowing they did not have to trap on those first two. We considered it 'practice bleeding'. When naval aviators are asked what it is like to land on a carrier, the most common response is that it is 'a cross between having sex and being in a train wreck.' Having accumulated nearly 300 traps, I would say that the day landings are closer to the sex, and the night landings are closer to the train wreck. There is a good reason why you only do day carrier landings prior to wings: retention. Night carrier landings are, without a doubt, the extreme test of aviation skills. All aspects of flight, glide slope, line-up and angle of attack, must be right on. The hook clearance on the F-14 over the round-down is about 14 feet. Line-up tolerance is just as critical. If your speed is fast, the hook won't catch (nose low, tail up) and if you are too slow, you risk an in-flight engagement, slamming the aircraft to

the deck. Do this all at night, in weather and with no diverts within a thousand miles, and you not only have a test of skill, but one of character.

"Preparation for carrier landing is a long process. We practised night after night on a specially lit runway. We used San Clemente Island, a good training site due to the lack of lights from a city or base, and the fact that there was a 60-foot cliff just short of the runway. 'Work-ups' didn't end after your initial qual, as landing on the boat is a fleeting skill. The need to stay 'current' is ongoing, and there are various safeguards in place to ensure that pilots stay 'in-qual' once on the ship. These rules require day traps prior to night traps, and/or a divert field should a pilot go for more than seven or fourteen days without a carrier landing. While shore-based, we constantly went out of qual and had to work back up to a proficient level prior to deploying. This applied to pilots with 2000 traps as well as to those with ten. It's *not* like riding a bike.

"During work-ups and refresher quals I was on the flight deck when a good friend's career came to an end. Bhudda and his RIO, Boog, were out for refresher night qual during work-ups for a cruise when he pulled off a little too much power for a little too long. The F-14 has a slow 'spool-up' time for the engines, and it was too little, too late. He hit the round-down with the tail and the hook of the airplane. He continued down the deck and once again became airborne as I saw a large piece of the nettings steel support rattle by. The back of the plane was engulfed in flame and he was instructed to eject. Luckily, they both got out and were uninjured. They were picked up by the helo within minutes. Bhudda never made it back. He was given all the help he needed, but could never get past the fear of night traps and, eventually, quit the Navy. He had excelled for two years of arduous training, was one of the best in all phases of flight, and yet, due to one mistake, he never made it to the fleet.

"Tom 'Boog' Powell became my best friend during my time in the Navy. Initially, we didn't get along well on the cruise while sharing the junior officer's bunk room. Personalities sometimes clash when nine guys share a small room. Later though, we did most everything together. We rode bikes, played golf and went skiing every January while on shore. We even took our bikes on the boat, to ride while on port calls. Boog was a RIO. He had stayed in the Navy Reserves and continued to fly the F-14. He was flying a training mission and his pilot lost control of the plane. The pilot was unable to regain control and they were forced to eject. This was the second time Boog had had to eject, and he was killed when he hit the canopy on the way out. The pilot was uninjured. Carrier aviation is so unforgiving. I miss him.

"Instrument flying is the key to carrier operations, and carrier pilots are the best instrument pilots in the world by necessity. There are no lights in the middle of the ocean, and when the skies are overcast or there is no moon, there is no horizon. More than one pilot has flown into the water at 300 knots, never realizing that he was disoriented. Due to the nature of their operations, Navy pilots must learn to handle vertigo and trust their instruments implicitly. Their lives depend on it.

"Better dead than look bad at the boat. A common phrase among naval aviators. Landing on the boat is what counts. It is a specific skill related to, but not dependent on, a pilot's abilities in other phases of flight. We had one F-14 pilot who would not take his plane in the vertical during air combat manoeuvring. Obviously his 'fighting' skills were limited to weapons and speed. But he could land on the boat. Therefore, his skills were preferable to those of a pilot who could out-manoeuvre the best fighter pilot, but who had trouble getting aboard the boat. Operating on the boat is what it's all about. Hardly anyone sees you shoot down your opponent, hit the bullseye, or fly perfect formation, but they all see you land

on the boat. The pilot's and, more importantly, the squadron's reputation is always on the line. There is a camera system that shows a live TV view of the flight deck from several angles, including a camera mounted in the flight deck on the centre line. There are TVs in all of the ready rooms and throughout the ship for anyone to monitor. There is also a requirement for a squadron representative to be with the Air Boss during day ops, and in PriFly at night, to answer any question that may be raised about a squadron mate's performance. Most of the questions are rhetorical and the red-faced squadron rep makes a logbook entry about which the pilot/guilty party must respond to later.

"The hook found a 'wire', and the trap now led to another under-appreciated challenge: taxiing on the flight deck. The interval between landing was typically about 45 seconds, so it was essential to clean up and get out of the way quickly. Once your aircraft stopped, the cable pulled it back a few feet and released your hook, allowing it to be raised. At the same time, you are retracting the flaps, sweeping the wings, and looking for the signals of the taxi director yellow-shirt. Taxiing on the flight deck can sometimes be uncomfortable if the deck has become slippery from oil, fuel, salt spray, etc. Combine a slippery deck with heavy seas and at times your airplane would slide. Usually there was the need to swing the nose out over the side of the deck in order to squeeze all of the aircraft onto the deck. Sitting 70 feet above the water as the deck pitches down, and feeling the wheels bump the scupper, made for a very uneasy feeling. Once out of the airplane, I usually sought the quickest route to the 03 level, just below the flight deck and out of harm's way. The flight deck is a very unsafe place to work. More than one unsuspecting sailor has walked directly into a whirling propeller or been sucked down the intake of a jet. The noise level is so high that you often can't tell where the threat lies."
– Frank Furbish, former US Navy fighter pilot

The small group of pilots who make up the LSO team must be in constant sync with each other to be certain that the controlling LSO is kept aware of the condition of the flight deck, clear or fouled. Safety is always the primary consideration in all flight deck operations.

The ingenuity on the boat was at times entertaining. We had an F-14 that lost its snubber pressure. That is a nitrogen pre-charge that puts down pressure on the hook to prevent it from bouncing back up after it hits the deck. "Mullet", the pilot, made several passes and on each attempt, the hook skipped over the wires (bolter). Then a chief appeared with several rolls of toilet paper which he positioned under the number three wire to raise it higher off the deck. On the next pass, Mullet was aboard.
– Frank Furbish, former US Navy fighter pilot

AIR STRIKE VIETNAM

REAR ADMIRAL Paul Gillcrist, US Navy (Retired) was Commander Gillcrist in the spring of 1968 when he was flying F-8 Crusaders from the USS *Bon Homme Richard* (CVA-31) in VF-53. Here is his description of his most memorable mission, and one man's heroism.

"How could I ever forget him? He did a very courageous thing . . . putting himself in great jeopardy . . . and for me . . . and he did it with such casual grace that contemplating it now, thirty-one years later, still makes my thoat constrict.

"It was a rather pleasant spring day . . . sunny skies, balmy breeze, blue sea and a bluer cloudless heaven. What more could a carrier pilot ask? Then, of course, there were those god damned sea snakes. As I walked forward on the flight deck towards my airplane, I looked over the starboard catwalk at the surface of the Tonkin Gulf. It almost made me sick. As far as the eye could see there were hundreds of thousands of sea snakes sinuating through the water in clusters of a dozen or so. They ranged in size from about two feet to five feet. Of a yellowish green colour, they swam just below the surface of the water with only their heads sticking out. Our intelligence officer had briefed us on them before we arrived in the gulf for the first time and told us they were the most poisonous of all reptiles on the planet. The thought of parachuting into the water filled with those hideous things made my stomach churn.

"Our flight headed inbound to a highly defended target in the Hanoi area of North Vietnam, and the year was 1968. The mission was photographic reconnaissance to assess the damage that had been done by a strike just thirty minutes earlier. Since it was such a highly defended target, I made the decision to go in as photo escort armed with four vice two Sidewinder air-to-air missiles. Our normal load-out was two of the deadly missiles, one each mounted on a single pylon on either side of the airplane's fuselage just aft of the cockpit.

The reason for this was aircraft weight. Our tired old F-8E Crusaders had grown in weight over the years from structural beef-ups to the addition of electronic warfare equipment and deceptive electronic counter-measures devices. The weight had become a problem, since each additional pound of gear meant one pound of fuel less with which we were allowed to land back aboard ship.

"At first it didn't seem to matter much . . . nothing more than a minor operational restriction with which we had to abide. But gradually, as the airplane's empty weight increased, we began to realize that our number of landing opportunities was decreasing by one for every 200 pounds increase in empty. At night a landing used up three times as much fuel. So, for every 600 pounds increase in the empty weight of the airplane, one less attempt to land was imposed.

"Back in the Pentagon in 1974 I did a study for my own personal interest. Using the A-4, F-8, F-4 and A-7 as examples, I found that, on average, a Navy tactical carrier airplane grew in empty weight at the rate of one pound per day of operational life. The accuracy of that rule of thumb was startling. In the case of the F-8, for example, the increase in weight was due to structural beef-ups, the addition of electronic warfare capabilities and electronic warfare counter-measures.

"So it was with careful consideration that I opted for the four missile configuration because, after all, we were going into MiG country. Who knows? I might need the extra two Sidewinders. As was the standard practice, the photo pilot, flying an RF-8, took the flight lead. I was his escort and his protection should the North Vietnamese decide to send out their MiGs after him. We 'coasted in' about twenty miles south of Haiphong, going at 'the speed of heat' as usual, and I found that, because of the drag associated with the extra missiles, I was using a lot more afterburner than the photo plane.

"The target, which was a railroad bridge between

Right: A Douglas A-4 Skyhawk, or Scooter, as it was called by US Navy airmen during the Vietnam War, ready to launch from the USS *Kitty Hawk* (CV-63) which was operating on station with the 7th Fleet in the Gulf of Tonkin off the coast of Vietnam. This example, a light-weight attack version, bears five bomb racks capable of carrying twenty separate items of ordnance weighing up to a total of 8200 pounds.

Above left: A Skyraider of VA-25 during the Vietnam conflict. Below left: In the Tonkin Gulf a Skyhawk aboard the USS *Midway* (CV-41) is ready to be launched. Above right: Captain James O'Brien, commanding officer of the *Midway* in 1965. Below right: A-1 Skyraiders of VA-25 on the *Midway* are being loaded with a variety of heavy ordnance for a raid on North Vietnamese military positions in 1965.

Haiphong and the capital city of Hanoi, was heavily defended . . . we already knew that. Nevertheless, the quantity of flak was startling as it always is. They were waiting for us, knowing that it was US policy to get bomb damage assessment after each major raid. I was flying a loose wing on the photo plane, scanning the area for MiGs of course, but also for flak because I knew that Ed would soon have to bury his head in his photo display shroud for the final few seconds of the run to be sure that the bridge span was properly framed in the camera's field of view. Neither he nor I wanted to mess this run up and have to come back another time. That would be too much. During the actual picture taking part of the run, when the photo pilot was too occupied to observe flak, I made it a habit to be in a strafing run at the most likely source of flak in the vicinity of the target. It always made me feel better, since just sitting there being shot at is always unpleasant and unnerving.

"Just as Ed settled down for the photo portion of the mission, which took an eternity of about five seconds, I heard a low SAM [Surface to Air Missile] warble (radar lock-on) followed immediately by a high warble (SAM launch) and my heartbeat tripled in an instant. I was on Ed's left flank and saw the SAM lift off at about four o'clock and only maybe ten miles away. There was a huge cloud of dust around the base of the missile as it lifted off and levelled almost immediately as it accelerated towards us. There was nothing I could do but call it out knowing that Ed would have to break off his run at the very last minute. (I think the North Vietnamese knew precisely what they were doing.) There was no way Ed could stay in his run because the missile was accelerating towards his tail at a startling rate. My mouth was dry as I keyed the microphone.

" 'One from Two. SAM lift-off at four o'clock, ten miles. Break hard right now!' Ed broke and I did the same, feeling the instant onset of at least nine

g's squeezing the g-suit bladders on my legs and abdomen. The SAM roared on by us just as another high warble came on in my headset. This one I didn't see and that really bothered me. 'No joy on the second one', I shouted into the mike. 'Keep it coming right, Two.' Then I saw it and it scared me badly since it was now off my left wing and coming at us fast. 'Reverse it left and down', I shouted hoarsely over the radio. Watching it pass to our left and explode about 200 yards away, I saw that we were skimming the treetops and called out to Ed: 'Two, let's get out of here!' I heard two distinct clicks of a microphone and knew that Ed agreed with me. The water was only a few miles away by this time and we were headed straight towards it. The two Crusaders thundered across the beach at perhaps 100 feet in full afterburner and doing in excess of 600 knots.

"About ten miles off the beach, when we knew we were outside the SAM and flak envelopes, we came out of burner and commenced a climb. Ed's voice came over the radio sounding apologetic and sheepish. 'Two, from One. I missed it.' We both felt badly. Bomb damage assessment of the bridge from the previous strike (half-an-hour earlier) was important and we both knew it. Another strike would be launched within the hour to go after it if the bridge span was still standing. As we passed to the seaward side of the northern SAR destroyer, I suggested that we report in to the ship's strike operation centre and ask them what they wanted us to do — although I was already fairly sure what the answer would be. We orbited at 15,000 feet over the Gulf as Ed checked in and reported failure, and asked for directions since the strike seemed important to them. Panther Strike told us to wait while they checked with Task Force 77 for further directions. As suspected, after waiting what seemed like an eternity, they told us to go back and try again. After the photo run had been completed we were directed to return to the ship as fast as we could so that the results could be

The Vietnam war drags on / In one corner of our living-room. The conversation turns / To take it in. Our smoking heads Drift back to us / From the grey fires of South-East Asia.
– *The Newscast* by Ian Hamilton

One minute we was laughin', me an' Ted, / The next, he lay beside me grinnin' — dead. 'There's nothin' to report,' the papers said.
– '*Nothing To Report*' by May Herschel Clarke

Boys are the cash of war. Whoever said / we're not free-spenders doesn't know our likes.
– from *New Years Eve, This Strangest Everything* by John Ciardi

Left: A US Navy air-sea rescue helicopter crew about to depart their carrier in the Gulf of Tonkin in 1966. The aircraft is believed to be an SH-3 Sea King. The Sea King replaced the Navy plane-guard helicopter, the Kaman UH-2 Seasprite then in wide use. The Sea Kings were deployed aboard carriers in three main versions, for plane-guard, anti-submarine operations, and combat rescue and recovery.

analyzed and another strike launched if necessary. At my urging, Ed requested that an airborne tanker be made available for us upon our going feet wet . . . and back we went. And they were waiting for us.

"Ed made a similar approach, using a different ingress route, and this time, flak and a SAM warning notwithstanding, we got the pictures. We both overflew the bridge in excess of 600 knots and made the turn towards the Gulf via a pre-planned route that ran on a southerly heading just west of Haiphong and then almost due east to the water over a relatively unpopulated area. Just as we approached the target my low fuel warning light illuminated and it shocked me. Nonetheless, we continued our egress from the target area as planned and went feet wet low and at high speed.

"I was elated because I was sure it was a successful photo run and neither of us had been hit. This time there hadn't even been any SAMs launched, just a radar lock-on. Ed had plenty of fuel remaining and began his climb at full power with the intent of hurrying back to Panther. I kissed him off and, looking at a fuel gauge which read only 400 pounds, I headed for a rendezvous with the tanker who had dropped down to 10,000 feet twenty miles south-west of the northern SAR destroyer. My mouth was dry as I began my rendezvous with the tanker. With only 400 pounds of fuel my engine would flame-out in just 15 minutes. This was, I thought ruefully, cutting it too god damned close. The tanker had been refuelling some A-4s in the air wing and been vectored to the northern SAR destroyer by the ship when we asked for a tanker before going back in for the second run.

"Naturally, I was extremely happy to see the tanker and was equally anxious to get plugged in. When I told the tanker pilot my fuel state his voice suddenly sounded a little strange. He was extemely apologetic and explained to me that he had just given away all of the fuel in his buddy store. I was shocked. How could this have happened? Somebody back at the ship had really screwed up by failing to tell him to hold at least 1500 pounds for me. A cold chill crept over me. There was no way I could get even halfway back to the ship with only 400 pounds of fuel.

"It is worth spending a few words of explanation on fuel and what it means to a carrier pilot. In any fleet squadron there is SOP (standard operating procedure) which dictates landing back aboard ship with a reasonable fuel reserve to take care of emergencies like a crash on deck, bad weather, malfunctioning recovery systems or a recovery delay for any of a dozen other reasons. For example, the low fuel level warning light comes on in the cockpit of an F-8 Crusader at 1100 pounds remaining. No carrier pilot who wants a long, safe career should ever be caught airborne with that light illuminated. There is a similar light in the A-4 which comes on at about 1100 pounds. So to find oneself 125 nautical miles from the ship in a Crusader with 400 pounds of fuel is not just critical . . . it is way past critical . . . the stuff of which nightmares are made, which still makes one wake up from a sound sleep with sweat dripping off one's forehead.

"By now we had joined up and were climbing to cruise altitude for our return to the ship . . . except I was not going to make it. It was a strange feeling . . . one of finality. It was all unreal. And again I thought of those sea snakes. We levelled off at 20,000 feet, and with my fuel gauge now reading 100 pounds I began to prepare myself for ejection. We were flying close together. The lower my fuel reading the closer I flew to him. Perhaps proximity to a friendly face made me feel comforted.

"The A-4 tanker pilot throttled way back to match a maximum endurance profile for the Crusader. Since getting back to the ship was out of the question, we were buying me some time before I flamed out . . . just a few more minutes . . . but every minute now seemed very precious. The tanker pilot's head, which was only 50 feet away from mine, turned to look at me for what seemed like a long time but was probably only 20 seconds. Then I saw him look down in the cockpit. A moment later the propeller on the nose of his tanker buddy store began to windmill in the airstream. This was the driving mechanism which reeled the refuelling basket in and out.

"The buddy store consisted of a fuel tank, a refueling hose, a take-up reel and an air-driven propeller to deploy and retract the basket. The fuel tank contained 300 gallons (2000 pounds of fuel) when full. The tanker pilot also had the capability of transferring fuel into and out of the tank from his own internal tanks. It was the tanker pilot's responsibility always to retain enough internal fuel to get back safely to the ship. I found it curious that the propeller was turning and then was startled to see the refuelling basket reel out to its full extension. The pilot then looked over at me and transmitted over the radio.

" 'Firefighter Two Zero Four, I have just transferred 500 pounds of fuel into the buddy store. Go get it.' I couldn't believe my ears.

" 'What about you?,' I asked, almost afraid to do so.

" 'You'd better take it while you can,' he warned ominously as though he were already having second thoughts. I slid back into position and prepared to tank. Never was my tanking skill so necessary as this moment. There could be no missed attempts. My last glance at the fuel gauge showed 100 pounds. I resolved not to look at it again. The sight made me physically ill.

"The in-flight refuelling probe on a Crusader is high on the port side of the fuselage just aft of the pilot. In actual measurements the tip of the probe is exactly 31 inches to the left of the pilot's eye. Therefore the inner rim of the refuelling basket, when the probe is centred in it, is only about a foot away from the canopy. The basket seems to float around like some wayward, feathery entity whenever one tries to engage it with the probe. In actual fact the basket, for all its airy movements, is in the tight grasp of a 250 knot gale. The airstream grips it like a vice. Thus it has all the resiliency of a steel rail. If it so much as touches the canopy of a Crusader, the result is an instant and violent implosion and fragmentation of the plexiglass into a thousand tiny shards which end up everywhere inside the cockpit. To say that tanking is a touchy evolution is the understatement of the year.

"Needless to say, my technique on this particular tanking attempt was flawless and the tip of the probe hit the basket in its dead centre with a 'clunk' which resulted in a small sine wave travelling up the hose to the end of the buddy store. It is a comforting sound and sensation which one can feel in his pants and right hand. It is the next best thing to sex.

"The very act of tanking, of course, uses fuel. It therefore took me about 100 pounds to get the 500 pounds, which left me with a net gain of 400 pounds. Now my fuel gauge read a much more comfortable 500 pounds leaving me feeling in 'hog heaven'. I disengaged and slid once more out to the right side of the tanker. We looked at each other and there was an unspoken understanding in the tilt of the tanker pilot's head that he had bought me some time . . . perhaps only fifteen minutes . . . but a very precious increment of life nonetheless. Experiencing no small amount of guilt, I felt compelled to ask him the obvious question.

" 'Can you make it back?' The answer was delayed.

" 'I'm not sure,' he said, but quickly added, 'There's another tanker up here somewhere. Maybe he's got a few pounds to give.' The tanker pilot then inquired of the ship about the tanker that had been sent north to tank the BARCAP (barrier combat air patrol). The ship came back quickly and informed us that he was indeed returning at maximum speed to rendezvous with us. The voice which told us that was more mature and sounded more senior. I

No man who witnessed the tragedies of the last war, no man who can imagine the unimaginable possibilities of the next war can advocate war out of irritability or frustration or impatience.
– from an address to the United Nations General Assembly on 11 November 1961 by President John F. Kennedy

War involved in its progress such a train of unforeseen and unsupposed circumstances that no human wisdom can calculate the end. It has but one thing certain, and that is to increase taxes.
– from *Prospects on the Rubicon* by Thomas Paine

War is cruelty, and you cannot refine it.
– from *Memoirs* by William Tecumseh Sherman

Let him who does not know what war is go to war.
– Spanish proverb

suspected immediately that, recognizing that they had screwed up this matter royally, they had put the first team on the problem. On a carrier at sea, everyone is in training for the next higher notch on the ladder.

"We droned along, still at maximum endurance speed, watching our fuel gauges and the distance measuring device on the TACAN display. It was really simple mathematics, the kind naval aviators learn to do quickly in their heads. My fuel flow gauge told me the engine was burning fuel at the rate of 2400 pounds per hour (divided by 60 converts to 40 pounds per minute.) My airspeed

indicator told me my speed was 300 knots. Indicated airspeed decreases as altitude increases at the rate of two per cent per thousand feet. Therefore, at my altitude of 20,000 feet my true airspeed would actually be 40 per cent higher than what my airspeed indicator read (40 per cent of 300 equals 120 which, when added to 300, equals 420 knots true airspeed. Divide 420 by 60 and it comes to 7 miles per minute. If I am burning 40 pounds per minute, then each mile I traverse through the air costs me six pounds of fuel.) By cross-checking the distance-measuring equipment reading on my airspeed indicator telling me how far away the carrier is, I corroborate that my ground speed is roughly what my true airspeed is, meaing little or no wind effect to worry about.

"But, no matter how often I ran the numbers through my head, they told me that my airplane would flame-out before I got to the ship. Again I thought of those god damned sea snakes. Our only hope was that the other tanker had some fuel left to give. Moments later a target appeared on the left side of my radar scope 30 miles away and converging. A few moments later, my tanker pilot, who had eyes better than mine, called out somewhat excitedly, 'Tally Ho the tanker ten o'clock fifteen miles.' The several minutes it took for us to complete the rendezvous seemed like forever, during which we were informed that there was 1400 pounds of 'giveaway' fuel available. My tanker pilot said, 'Firefighter, you go first and take 800 pounds and I'll take 600, okay?' What could I say? My fuel gauge again showed 300 pounds.

"Of course, the ship gave us priority in the landing sequence. We made a straight-in approach with the tanker taking interval on me at about ten miles astern. The sight of the ship steaming into the wind with a ready deck is one of the most beautiful sights in all of my memory. The LSOs seemed to understand our sense of urgency. We both caught the number two wire and taxied

forward for shutdown. The arrestment felt great, the feel of the ship under my wheels felt wonderful. The blast of warm air that filled my cockpit when I opened the canopy tasted like pure oxygen. Life was beautiful at that moment. As I climbed out of the cockpit I felt a sudden and enormous exhaustion. Then I took one last look at the fuel quantity gauge. It read 100 pounds.

"On the way from the airplane to the ready room I took a short detour to the edge of the flight deck and scanned the sea for the snakes. Strangely, there were none to be seen.

"Five minutes later, while finishing filling out the maintenance 'yellow sheet' in my ready room seat, I took a sip of steaming hot coffee and relished the moment. Then I got up, walked over to the duty officer's desk and pressed the lever on the 19 MC squawk box. 'Ready Room Four, this is Ready Room Two, Commander Gillcrist calling. Is the pilot of Four One Four there?' The answer was immediate.

" 'He's listening' said a voice.

" 'Young man,' I said. 'You have cojones of brass. I owe you one.' The now familiar voice of the tanker pilot came right back.

" 'No problem, Skipper, glad to be of help. It's always a pleasure to pass gas to a fighter pilot. I'll collect at the bar in Cubi.'

" 'You're on,' I finished.

"I never got to pay off the debt. Three days later, that fine young man was literally blown out of the sky by a direct hit from an 85mm anti-aircraft shell.

"The three salvos from the Marine honour guard's rifles jolted me in much the same way that the violent 85mm explosion must have done when it ended that young man's life. For the first time in my life, as I stood at attention, saluting with the bugler playing taps, the tears coursed down my cheeks. The flight deck was heaving slowly in mute response to a gentle swell. Then the Marine honour guard tilted the catafalque up. The coffin slid out from under the American flag and fell into the sea."

Above left: Aircraft mechanics with their charge, a Chance–Vought F-8 Crusader fighter in 1974 aboard the US carrier *Coral Sea* (CV-43). Left: An A-4C Skyhawk at the launch position on the USS *Independence* (CVA-62) in the Tonkin Gulf, 1968. Above: Paul Gillcrist as a navy test pilot at Naval Air Test Center, Patuxent River, Maryland in 1959. Overleaf left: The EA-6B Prowler entered service in 1972 as the premier electronic warfare aircraft of the time. Overleaf right: The Grumman E-2 Hawkeye was first deployed in the Vietnam War aboard the USS *Kittyhawk* (CV-63) in 1965, replacing the older Grumman E-1 Tracer.

Left: On Yankee Station in the Gulf of Tonkin, the *Bon Homme Richard* (CVA-31) in the company of the USS *Sacramento* and the USS *Thomaston*. Below: Commander Paul Speer, commanding officer of VF-211, in conversation with Joe Shea, one of his pilots, about a May 1967 mission over North Vietnam during which each of them shot down a MiG. They flew from the *Bon Homme Richard*.

SEAHAWKS

VMAQ

FIGHTING 84

KITTY HAWK

100

FLELOGSUPPRON 24

DESERT STORM

67

CV67 · CVW

ARABIAN GULF

YACHT CLUB

JOLLY ROGERS

TOMCAT

Above: A selection of unit and organization patches worn on the flight suits of US naval aviators.

SHOO SHOO SHOO, BABY

The victories of mind, / Are won for all mankind; / But war wastes what it wins, / Ends worse than it begins, / And is a game of woes, Which nations always lose:
Though tyrant tyrant kill,
The slayer liveth still.
– *War*
by Ebenezer Elliott

Letter from a sailor's girlfriend:
Dear John,
I couldn't wait any longer, so I married your dad.
Love,
Mom

Right: For navy personnel and their loved ones, parting is a sorrow that is just part of the deal.

ALL BRANCHES of the military have their problems, but in peacetime navies suffer more in one way than the other services. No matter how dedicated a person may be, or how much he likes his job, nothing can take the place of being with the ones he loves, or make up for the prolonged periods away from them. It's a hard fact. Every navy has its requirements and commitments which have to be met. They have to take precedence over all other considerations. Lengthy work-ups prior to lengthy deployments are just part of the deal, and that isn't likely to change.

What looks like fun, excitement, challenge and adventure to a twenty-three-year-old naval aviator fresh from the Fleet Replacement Squadron and on his way to the Fleet, may look very different to a mid-grade aviator officer who has experienced two or three work-up cycles and deployments which, cumulatively, have taken him away from his wife and kids for half of the six years he has been in the service. Never mind the benefits, the travel, adventure, security, the satisfaction in serving his country, even the thrill of flying military aircraft which drew him to the Navy in the first place and which he has wanted to do all his life more than anything . . . even that may no longer seem quite so essential. Nature has kicked in and he's only human. And not incidentally, just how much understanding can one expect from a wife or girlfriend?

What is a navy to do? How is it supposed to retain the guy who has taken the training, served his obligation faithfully and efficiently, and must soon decide whether to stay in or leave this service that he has been a part of for several years? It can't hope to match the salary he can get on the outside. He'd be of no use to the Fleet if the Navy were to let him stay ashore permanently — navies go to sea and that's what he was trained to do. If even the flying has lost some of its lustre for him, at least compared to quality time with the family, where is the incentive for him to re-up? And what about that seat with the airline that may be waiting for him? Economies have a way of changing, and that seat may not be there the next time he looks. What should he do? Stay in? Get out?

"The hard part of navy life is leaving your family for six months. Everyone will say that, but it's equally hard to do the work-up process that we go through. We go out for two or three weeks at a time; we're in [port] for maybe three weeks, and then out for ten. It just disrupts the family life.

"I like e-mail. I want to know if my son fell and hurt his head. I would want to know it right away. I would be concerned the whole time until I got that e-mail saying he was OK. You feel like you stay connected with all the stuff that's going on. For special occasions like Christmas and some holidays, they sometimes do a real-time video thing, and now there are a lot of satellite phones too.

"The introduction of women on the ships was a huge thing. There were growing pains associated with it and it was a topic of discussion with just about everybody for a long time. But we've all seen that women can integrate into the environment, and it's not a big deal. We still have some old salts, people who have been in the Navy for a long time who are gonna resist. But the Navy is evolving and changing, and those people are being asked to leave after they've fulfilled their twenty years. The new airman who checks in on board, who only knows a navy with women on the ships, it's nothing to him. The ship's company, including the air wing, is about 6000 people now, of which about 10 to 12 per cent are women.

"You're always gonna have people that complain, but the way you combat morale problems is, you keep people busy. If they are involved in doing their jobs, they may not be happy that they're away from home, but they're gonna be content doing what they're doing as long as they know there is a goal. I think morale on this ship is fine. We have movies, satellite TV, e-mail, computers, satellite phones, video games, workout facilities . . .

it all contributes to improving morale. We all made a conscious choice to be here."
– Danny Vincent, USS *John C. Stennis* (CVN-74)

"I'm looking forward to the six-month deployment. I don't mind being out at sea. I'm young. It's not that I have nothing to look forward to if I were to stay on shore, but I think the cruise is a chance for me to get out and see things. I'm not married. It's a chance for me to save money, and a lot of time to think, so I don't mind it. The foreign ports are great. In the Gulf we hit Dubai and Bahrain. On our transit we'll most likely hit Perth, Hobart, Hong Kong, Singapore and Hawaii."
– Aaron Zizzo, USS *John C. Stennis* (CVN-74)

"The pilots in the squadron, and there are only eight of us at the moment, get on very well indeed. In fact, there are only about forty to forty-five Sea Harrier pilots in the Royal Navy currently. We all know each other and we all get on. They are all like-minded people doing the job for fairly similar reasons. The cameraderie is great."
– Nick Walker, 801 Squadron pilot, HMS *Illustrious*

"I think the Navy is a great opportunity. Great people. I know I sound like a recruiter, but it's true. The caveat is, it's what you make of it. If you want to be miserable, you can be miserable. It's not hard to do. If you don't want to grab the opportunities, you don't have to, but they are there and, if you love your job, you'll find those opportunities. They'll find you, and you'll make the best of it and have a great career."
– Shannon Callahan, NAS Pensacola

"My fiancé and I have been very lucky. We met five years ago when we were both sophomores at the Academy. We were both lucky to get our first choice of aviation. I've been right behind him in the programme, about two to three months. We both got Whiting Field. We both got helicopters.

He got winged on 24 September and I got winged on 17 December last year. And we both ended up getting San Diego. The Navy has been great. We were engaged in October and by the time my selection sheet came around, I was able to put down that I was engaged to a guy who was flying in the helicopter community in San Diego. I put San Diego as my first, second and third choices, and got my first choice. Washington DC and the Navy have been wonderful in that we are both going to be in the same place for three or four years, and after that we'll be able to be co-located to our next duty station as well.

"Flying wasn't my first choice in the beginning, but from day one in the cockpit of the T-34, I've absolutely fallen in love with it. The missions we do are just incredible. Helicopters are more than I ever thought I could dream of flying. Obviously I haven't been to the Fleet yet so I don't know what it has to offer, but the places I've been, the people I've met and the things I've done have really given me a positive outlook on the Navy to this point. I can definitely see myself going for a career. I'm from a long line of teachers and I know that I would really like to come back, either to the FRS or to Whiting, and be an instructor. If that meant I could go as far as being XO [Executive Officer] or CO [Commanding Officer] of a training squadron, that would be wonderful."
– Michelle Vorce, NAS Whiting Field

"Separation and how to deal with it was an individual dilemma. At times, the feeling of isolation was enormous. I assume that the problem has been diluted with the advent of e-mail. During my time, the C-2 Cod [Carrier Onboard Delivery aircraft], lovingly called Mrs Piggy, was our infrequent connection to home. It would deliver the mail, which was obviously not timely but was always received with much anticipation. It was a gloomy day when you didn't get a single piece of mail or a 'care package' from home. Care packages were

Above: Captain Mark Stanhope, skipper of HMS *Illustrious* in 1999. Top and right: The joy of Navy homecoming.

boxes containing stale cookies or other treats. Besides the contents, these packages commonly contained portions of hometown newspapers used to wrap the food items and these papers would be distributed around the ready room. It was the simple things that were missed.

"Separation from family was the hardest aspect of ship life for most. Several members of the squadron came home from a cruise to be introduced to a new child. Then there were those who came home from a cruise to an empty house. I remember counselling an enlisted man who had just received a letter from his wife. It was a common practice to sign a power of attorney allowing a spouse the authority to run the household. In this case, the lady had used the power of attorney to sell his house and wipe out his finances. She then moved in with a friend of his. We had only been gone for three weeks. Life on the ship can be tough. There were others who revelled in the separation — an extended break from a bad marriage or a timely excuse to end a relationship. Most immersed themselves in their work, wrote and received loving letters, and returned to rebuild their relationships with their families."
– Frank Furbish, former US Navy fighter pilot

"The six-month deployment for me is both interesting and kind of scary. I have a wife at home with a pre-existing medical condition, so I have that in the back of my mind. If I was single or had a wife in healthy condition, I'd be looking forward to it. For her sake, if I had a choice, I wouldn't go on the deployment. We do keep in touch by e-mail, and my division are aware of what I'm dealing with at home. Once in a while they'll allow me to call home just to double-check on her and make sure everything is going OK.

"My job is very good and very rewarding. My father was an E-8 in the Navy, and from watching him in his career, taking on board a lot of the stuff that he did, I learned from him. Being in the Navy has

taught me about responsibility. I grew up real fast."
– Parnell Chapell, USS *John C. Stennis* (CVN-74)

"Joe Deigh was our Division Officer. [In addition to] his other duties, he was responsible for censoring the outgoing mail. One time I wrote a letter to my wife Mae and enclosed two sticks of gum. I explained that one stick was for her, and the other one was for the Censor. When Mae finally opened the letter she found just one stick of gum and a message from Joe himself: 'Thanks for the gum. Signed: the Censor.' [To me,] he was the best Division Officer aboard [the] ship."
– Mark Roses, formerly assigned to the USS *Enterprise* (CV-6)

"My girlfriend is very supportive actually. She has always been very supportive and she knew that I wanted to fly when we first started going out together. She's never tried to say it's too dangerous and why do I want to do it? She's always put up with me being away a lot and only seeing me at weekends, and now, of course, she's putting up with me being away for longer periods on the ship and certainly next year in the Gulf, for quite a long time. I think she's very pleased that I'm doing what I want to do and I'm very lucky to have someone who thinks that way. E-mail has made a big difference, although the number of e-mails you can send is limited. I think it's five or six a week sent to a private address. It's a fantastic way of keeping in touch and really, in this modern world, there's no reason why we shouldn't have it available. But even though we have got e-mail, receiving and writing letters I still find very important, and often receiving a letter is more personal. Someone has actually taken the time to do it, and it's nice to receive. I actually do write home quite a lot on snail mail as well."
– Nick Walker, HMS *Illustrious*

"There were always one or two members of the

squadron who were on the 'wild' side on shore leave, but most were the best guys in the world. You flew their wing in formation, or they yours. This took trust. You flew with your team daily, lived with them, lifetime friendships were formed. All of my squadron friends who survived the war later married and raised families."
– Richard H. May, former US Navy fighter pilot

"It's still up in the air as to which ports we're gonna go to on this Gulf deployment. On the port calls, I'd say 80 per cent of the crew head off to find the closest pub or club. The rest of us go out and discover the country; get a taste of the culture and the food and the people. I like the adventure. A carrier is usually in port for three to five days. We are the representatives of the Navy, and a carrier pulling in is a pretty big thing to another country, especially a small country that doesn't have a lot of people. We try to keep a good reputation there so we are welcomed back the next time, and as long as everything goes well, we are more than welcomed back. So far, most countries have been very welcoming. With almost 6000 people, we practically take over the town that we go into, so it's important that we are on our best behaviour while we're there. On our last cruise, every port we went to, they just loved us."
– Dale McGhee, USS *John C. Stennis* (CVN-74)

From a conversation with Captain Mark Stanhope, Commanding Officer, HMS *Illustrious* in 1999:
  "We are recognizing now that one of the important requirements to continue the good will and motivation of our people is to ensure that they can talk to their home. You know, the old letter is still an important mechanism, but it's somewhat dated in this day and age of instant communications, and while we're around the UK, everybody wants to use their mobile [cellular] phones. We have a system on board here on daily orders. It says Mobile Phone State Green which

203

means that they can use their mobile phones should they wish to. There are about 700 mobile phones on board this ship. Most people have standard mobile phones that you can use around the UK, but there are some mobile phones on board which are the new 'phone anywhere in the world' phones, and they like that sort of access. Away from home though, we've got to look at other mechanisms to keep them in touch. E-mail now is a great facility on board here, and they are allowed to send up to five letter-page e-mails a week.

"We have an operational requirement to have phone connectivity with headquarters, the Ministry of Defence and such like, which carries with it, on a ship of this size, about nine or ten different telephone lines. During the working day these are being utilized by various officers to do their business, but in the evening we take three, sometimes four of those lines and pass them through a phone box down below. The ship's company can go into this phone box and ring a number that gives them access to any net that they want to use to phone home. This costs. I think it's about 33 pence a minute. It's quite expensive. In my time in submarines, we went away for eight weeks and didn't speak once to family. Here, they have it available to them at cost. It carries with it a burden. I can't deny that. The burden it carries with it is that it has changed communications. The average sailor on board suffers the ups and downs of home life . . . little Johnny is sick, or the wife is ill, or the wife has just smashed the car up, or mother is ill, and so all the woes that we were, frankly, cocooned from in the past, as you go away and leave your wife to sort out all the problems, go home at the end of a trip and help her sort them out . . . now it's on a daily basis. So lots of information gets fed to people that they want to do something about but can't.

"We have something called the Naval Families Services System, which is an organization where, if there is a problem at home, an illness or a bereavement, or a marital problem, or anything, the wife can get hold of this organization. The organization can then go and see the wife and see just how serious the problem really is. They will then signal me with their recommendation. They will tell me what the problem is and maybe give me a recommendation that, in their opinion, it's important that we give this sailor three weeks leave to go home and sort out the problem, or support the family, or look after the children while the wife is ill, and I would normally react to them positively. It would only be an operational expedient that [would cause me to] say 'I'm sorry. I understand the situation, but I cannot spare this man.' That's finally my judgement. In and around the UK we would make that judgement fairly easily. You can get people on and off. What we can't do abroad is get them backwards and forwards very easily. When we go to the Gulf, for instance, it's easy for me to reduce the number of people on board this ship when we're doing an exercise. I'm not quite so free and easy about giving up one of my two radar maintainers when radar is critical to our operational capability. So that becomes slightly more demanding. How do I ensure that the ship's company can appreciate the difference? How can I make sure that when we get out to the Gulf they are clear that this isn't quite the free and easy time that they've had before, and that I'm going to be a bit more restrictive in dealing with them? My answer to that was that I think you've got to be as open as you can with the ship's company about the business in the first place. Set it up. 'Put money in the bank.' That is, in circumstances when it is quite possible to let the people go, let them go. Then when you can't let them go, at least you can say, 'Look, I've got a track record here of letting people go as best I can when the circumstances permit. Right now, the circumstances do not permit. I'm sorry. Sod's law says it happens to you when we were operational, and it happened to Jimmy Smith over there when we weren't operational.' I think that's really the only way you can do it. They did join the Navy and the Queen pays them, and therefore, in the end, you've got to take the Queen's shilling, as they say, and do the business.

"I think it is essential today to provide our people with instantaneous communication. I'd go further. I know in the American Navy, on deployment, they sometimes even offer to the ship's company video links with home so they can see little Johnny growing up. The wife goes along to wherever it is that she's got to go to sit in front of the video camera, and he sits down here and chats with his wife for ten minutes, looking at her with the kiddies surrounding her. That's where we've gotta go. This is what people want. I think there is no question but that this is what we've got to provide, and I don't think there's any question but that along with it will come certain problems. But the problems, frankly, are small beer compared with the motivation factor that one would have of trying to somehow offset this one difficulty the Navy has — all the armed services have — which is so much against the modern way of life, which is separation.

"People don't like being separated and you can't show the stiff upper lip in true grit British style and go away for two-and-a-half years. It wasn't so long ago, not in my time in the Navy, but certainly in my lifetime, that the Navy used to go away for two-and-a-half years. You'd sail from Portsmouth, leaving your wife and kids on the jetty. Two-and-a-half years later you'd come back. Now we have formal limits on how long these ships can be deployed. We have harmony rules on how long the ship's got to sit in its own dockyard port. Theoretically the planners should look at keeping these ships in dockyard ports, which is Portsmouth for this ship, for something like 43 per cent of the year, so you aggregate your time, and 43 per cent of that time should be in

Portsmouth. You don't aggregate it over a year. You actually aggregate it over two or three years to compensate for the fact that some years you're gonna be away more than others. That's what we're working to. We don't allow ships to go away for longer than six months, or if they do, someone's got to explain why that is, and if that's the case then allowances are made for a foreign visit, to support the wives going out for that foreign visit, or provide interest-free loan support to wives going out for that visit. This is very much part of the way we are doing our business now and the more we can make the lifestyle of these people acceptable, the easier it is to recruit them, and probably more significantly, to retain them. Much of what I'm suggesting does cost money; there's no two ways about it, but it can be leeched on to the side of the operational requirement anyway. TV! It again seems small beer, but the Government in one of their last defence reviews are now trying to fit satellite televisions to all ships. It's quite difficult to do, actually. This big platform is a challenge in itself. I can switch on the TV and get CNN because of its operational importance. In the Gulf, in the ops room in my command seat, I can convert the screen to present to me an intelligence picture, flight deck cameras, or CNN, and in the Gulf CNN was an important mechanism because they pick it up so much quicker than most. We're looking at putting these satellite televisions on all ships."

"One of the most valued possessions I took away from my stint in the Navy was the cameraderie built with my squadron mates. Your squadron becomes your family while on the boat. The bunk room member's problems become your problems and an indivdual's joy becomes a collective joy. Things didn't always run smoothly in that small room housing the nine of us, but overall we got along well. I still keep in touch with several of my fellow squadron mates. Along with the cameraderie,

On his wedding night / They took him to war. / Five years of hardship. / One day he returned / On a red stretcher And his three sons / Met him at the port.
– *Sons of War*
by Samih Al-Qasim, translated from the Arabic by Abdullah al-Udhari

You love us when we're heroes, home on leave, / Or wounded in a mentionable place. / You worship decorations; you believe / That chivalry redeems the war's disgrace. / You make us shells. You listen with delight, / By tales of dirt and danger fondly thrilled. You crown our distant ardours while we fight, And mourn our laurelled memories when we're killed.
– from *Glory of Women*
by Siegfried Sassoon

You said 'May God go with you, Son / And may you soon come safely home / Remember that where'er you go / You'll never be alone' / Well I carried those thoughts with me, Mum / But when the aeroplane took me higher / I couldn't see the Angels, Mum And I never heard their choir.
– from *May God Go with You, Son*
by C. Wright

serving aboard the boat builds a sense of duty and pride in your country. Most folks don't take in 'the big picture' and just go out and do their job while serving aboard the ship. Defending the rights of your country and developing pride in a job well done is a reward in itself.

"Perhaps the greatest morale booster for most of us was receiving mail. Those with sweethearts or wives and family really needed reassurance that everything was OK at home. Even the act of writing was therapeutic and stabilizing, and helped offset our loneliness and feelings of futility. It was sad to realize that some crew members never received mail or packages, and those of us who did, often tried to share as best we could. While being away from home for such lengthy periods strengthened personal bonds for some, it undermined them for others."
— Frank Furbish, former US Navy fighter pilot

When a carrier returns to its home port after a long deployment half a world away, the final few hours before the great ship can be manoeuvred to her berth, tied up, and connected to the quay by gangplanks, can be the longest hours of the cruise. Thousands of wives, children, girlfriends, parents, relatives and friends anxiously press forward on the quay, straining to pick out their particular sailor amid the ranks of officers and ratings lining the edge of the flight deck.

As the ship is finally secured, there are shouts of sailor's names, and hundreds of waving arms and hands holding up home-made signs: IT'S A BOY, HEY, BOBBY!, CAN'T WAIT!, I LOVE YOU ED, and WELCOME HOME. The restless, expectant crowd has been waiting six months for the return of this vessel and now the final minutes of waiting are almost agony. By Navy tradition, the men of the ship's company who are the fathers of babies born while the carrier has been on deployment, are entitled to disembark before all others. The first moments of the homecoming, tearful and happy, are treasured.

There's still no letter . . . / In my troubled mind / I seek a reason, and quickly reasons find, / Indeed they tumble in, to be discarded / Each as it comes . . . It could be that You're very busy; missed the evening post; / Or else it's held up in the mail. A host Of explanations . . . Yet that gnawing fear / O'errides them, still keeps dunning at me that / You just don't want to write. / And vainly I Attempt to thrust aside the thought; deny / It with your last note, and the one before. / But no. I must resign myself to wait Until tomorrow, or the next day and / A day. Surely then I see your hand — / Writing and envelope. And life is sweet, until A week or so, when . . Still no letter.
— Still No Letter
by John Wedge

207

SLIDERS

NAVY PILOTS call it a slider. It's a burger, and a slider with cheese is probably the most popular order at the "dirty shirt" wardroom — the canteen on a carrier where pilots and NFOs in their flight suits are always welcome.

Kevin Morris joined the Royal Navy in 1977. He went in as a Cook's Mate and made his way steadily through the ranks. He went from Cook to the training to be a Chef. From that, he transferred to Catering, moving away from the "craft" side to management, including accounting, budgeting and stock management. In 1993 he attained the rank of Chief Caterer and was selected to go to Dartmouth and take a commission. As an officer he held positions at the Navy's Cookery School where he was responsible for training cooks and stewards, and another shore position in which he managed contracts and camp catering. As of October 1999 he has been aboard HMS *Illustrious* for one year, in charge of Food Services. His responsibilities include anything and everything to do with food: from the menus, to hygiene standards, to making sure the ship has enough food on board at all times. He is concerned with anything to do with the galley, the store rooms, the ship's company eating areas and the dining halls. Nothing to do with food and dining aboard *Illustrious* escapes his attention:

"Several inputs go into the menu planning, including budgetary input. It's no good putting something on the counter that people don't like or won't eat. The menus are constructed on a weekly basis, and we run from a framework of six menus on a rotational basis. Each week we look at the menu that is next in the rotation and make minor adjustments relating to stock remaining on board that we need to turn over. We also look at the ship's programme, where the ship is going to be in the world. We don't want to put on a hot, stodgy pudding if we're going to be in the tropics. We also have to consider what the ship is doing. We have to adjust our feeding methods to comply with the ship's programme. If we are to be at Action Stations, we generally have Action Messing. If there is a RAS (Replenishment At Sea) going on, we have to look at the style of menu that we put on and the timings of meals because a lot of people can't come down for a meal at normal times. If, for instance, we are taking on fuel, there are a lot of people involved in special duties throughout the ship who have to remain at their duty station and may not be able to get down for a specific meal. We have to make provision to feed them wherever they are stationed, or extend the meal times so they can come through afterwards. After we have looked at the programming side, we try not to be repetitious in our meal planning. But when you are putting four hot choices on for lunch and four hot choices on for the evening, eight popular hot choices a day, seven days a week, you quickly run out of popular choices. We have to make sure that a dish doesn't come around too often. We have to make the menus as varied as possible.

"Certain trends have been filtering through in recent years. There is a general move away from red meat consumption and towards fish and chicken. Vegetarianism is coming on in a whole range of forms. There are a number of tastes to take into account. We are also getting a lot of ethnic minorities coming through, which provides us with another challenge. Still, many of the dishes that we prepare and serve today are the same ones that we were serving ten to fifteen years ago.

"The biggest thing on board is morale. Food and morale are very closely linked. You can have the best machines, the best equipment, the best ships, but if you can't feed people well, you're not going anywhere.

"One of my biggest jobs is to make sure we have enough food, and for 1200 people, that is an awful lot of food, and you run out very quickly. Ideally, we need to replenish every two weeks. We go through sixty to seventy pallets of food a month.

"Certain items are very difficult to get outside of the

Left: Diamond Head and Waikiki Beach during World War II. Below: Recreation in the form of a friendly game of cards aboard a US Navy carrier in the Pacific during the war.

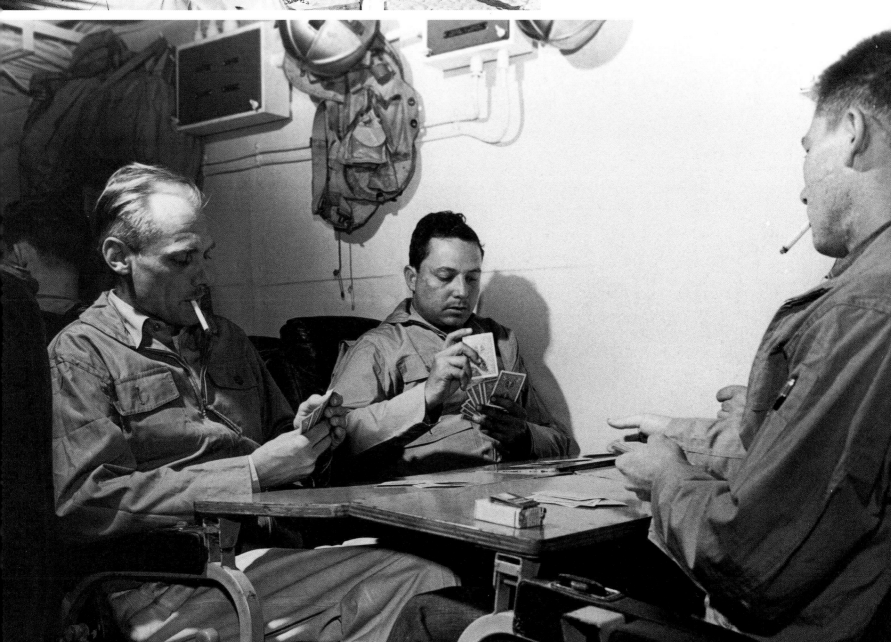

UK. Sausages are very popular; everyone has them for breakfast, and you just can't get them anywhere else. We just bought some in Amsterdam because we had run low, but they didn't go down well with the ship's company. It's one of those things."

Benny L. James was serving on the USS *Enterprise* in the autumn of 1944. "My brother came to visit me while we were riding at anchor in Ulithi in the hot Pacific. I was standing watch on gun mount number five when I heard my name called. I turned around and saw my older brother. We had not seen each other in three years. He was stationed on a destroyer, the USS *Halsey Powell*, which was anchored up in the north anchorage.

"He stayed all night and ate supper with me. We had a good visit and I showed him around the 'Big E'. We talked most of the night and slept awhile on the gun mount. He was supposed to be back at his ship by 0800 hours. As the time approached for him to go, we could not find anyone who was going in that direction. We asked several coxswain if they would take him to the north anchorage, and they all responded in the same way — they wouldn't take him for a hundred bucks. I said that it was only fifteen minutes up there, and they said they wouldn't do it for two hundred dollars. Finally, in desperation I asked a boatswain mate if he would take a gallon of gedunk [ice cream] for [making] the trip. He looked at me hungrily and said: 'make it two gallons and you're on.' I said I would throw in a pint of chocolate syrup and he about went nuts.

"When I went down to get the ice cream, the man who ran the freezer said that no one got any gedunk until after 1100, and then only the flyers and officers. When I explained the situation he relented and slipped me a couple of gallons in a box and said not to tell anyone for at least thirty years. I said, for a pint of topping, I would clam up for fifty years. When I got back up to the aft port crane they were ready to shove off — without my brother. When I finally got the boatswain mate's

210

attention, and let him know what I had, he motioned my brother [to come] aboard, ice cream and all. The last I saw of them, they were dipping in the ice cream and heading north at full speed. I later got a letter saying my brother was only fifteen minutes late for muster."

"Apart from flying, there were three big events of the day: breakfast, lunch and dinner. Not that the food was that good, just that eating was something to do, a popular diversion. The officers had two messes. One was on the second deck. It was more formal and run by the black shoes (naval aviators wore brown shoes). It was set up buffet-style and enlisted personnel assisted with the clean-up and beverage refills. There were specific protocols seemingly designed to keep the aircrew out. One was the requirement to wear khakis. Flight jackets were also forbidden. Aircrew wore flight suits, unless prohibited by having to stand watch or some other function. So having to change uniform in order to eat was not worth the effort. Thus the aircrew ate in the 'Dirty Shirt' located on the 03 level, just two knee-knockers from our ready room. The Dirty Shirt was also buffet-style and had a small eating area. There were no waiters. Eating on the ship can be compared with many other facilities that feed thousands, such as hospitals, colleges and airlines. Most of the food was mediocre. We had names for certain food items. Rollers are hot dogs and sliders are hamburgers. The reasons for the names become apparent during heavy seas. Trail markers are brussel sprouts, and bug juice is Kool Aid. Auto-Dog is not a food, but a machine for dispensing soft ice cream. Most guys would spiral the ice cream on a dinner plate, or lay it down in a straight line, hence the name. Another machine related to eating was the Cake Dryer, rumoured to have sucked all the moisture out of cakes and other baked products.
  "Entertainment was difficult on the boat. Fortunately, you were usually busy with flying,

collateral duties, watches, etc. The ship had a TV station that played movies, news and canned videos provided by contractors, relating to ships and aircraft. The most popular show was the 30-minute workout video shown in the morning, of women doing exercises to music, a tradition that has probably gone the way of the politically correct Navy. The Plat (ship's TV) was also constantly on to monitor flight deck operations. VCRs and computers were staples in the staterooms and in the ready room. Personal computers were life savers, not just for work, but also for entertainment. While on cruise, I taught myself how to program in dBase and wrote programmes to assist me in my collateral duties. We had a few computer games to pass around. The advantage of e-mail then would have made life seem much less isolated. Playing cards was another favourite diversion. Generally, there was a poker game running within our squadron several nights a week. Each new game was named after the last port call, and the stakes were small and usually friendly.
  "The ship provided several diversions. On occasion flight ops would be suspended and there would be a flight deck picnic. Grills were moved to the deck for a meal of steaks and chicken. Some would set up chairs and towels to bask in the sun and just relax. Others would jog. Once, my RIO and I were assigned to stand 'alert 5' during the flight deck picnic. One would expect that the alert would have been cancelled due to the presence of a thousand people on the deck, obviously preventing any launches. Wrong. We had a concert performed by Kris Kristofferson during one line period. The most legendary tradition on the ship occurs when we cross the equator. Pollywogs — people who haven't crossed the equator before — become Shellbacks and are given a certificate commemorating the event. The entire ship is transformed into a gigantic hazing exercise, with both officers and enlisted personnel utilizing leftover food, grease, hoses and water to

WALDORF SALAD
(yield: 100 portions)

Ingredients:
3 tbsp (3/4 oz) milk, non-fat, dry
7/8 cup water, warm
1/2 cup lemon juice
1/4 cup (2 oz) sugar, granulated
1 1/4 qt (2 lb 8 oz) salad dressing
3 qt (4 lb) celery, fresh, diced 1/2"
1 qt (1 lb) walnuts, coarsely chopped
2 1/2 gal (9 lb) apples, fresh, eating, unpared, cored, diced 1/3"
4 lb lettuce, fresh, trimmed

Method:
1. Trim, wash and prepare fruit and vegetables as directed on Recipe A-31.
2. Reconstitute milk.
3. Combine lemon juice, sugar and salad dressing. Add to milk. Mix well.
4. Add celery, nuts and apples to salad dressing mixture. Toss well to coat pieces.
5. Place one lettuce leaf on each serving dish; add salad. Cover; refrigerate until ready to serve.

Top left: Taking exercise on the flight deck of HMS *Illustrious* while crossing the Bay of Biscay in October 1999. Bottom left: Food service for the enlisted ratings of *Illustrious* in 1999.

Below: Chief Myers and
pilot "Dusty" Kleiss
enjoying a cool beer at
the Scouting Squadron
Six picnic, Barber's
Point, Hawaii in 1941.

unpleasant effect. Shellbacks count their 'graduation' certificates among their most prized possessions, in order not to have to do it all again. The best diversion though, is the next port call.

"Port calls are the light at the end of the tunnel. Morale peaks just prior to a port call and, of course, the end of the cruise. The PAO (Public Affairs Officer) sends in an advanced detachment to set up tours and hotel packages, and squadron officers also go in early to set up the admin and establish hotel accommodations. The admin was usually a suite which was the central meeting place for the squadron officers. It was run and maintained by junior officers who were allowed (required) to stay there and protect the booze locker and the squadron memorabilia (flag, mascot, etc.) Often, the advanced det would hand out flyers to any half-attractive female to come to a party held at the admin the first night in port. This never came off quite as well as anticipated. Apparently, spending too much time on the boat obscured and distorted one's vision and perspective on female beauty. Most guys would go to the party, put in their required 'face time', and then go elsewhere. Port time was normally spent in typical tourist activites, sightseeing, dining and visiting the 'hot spots'. Booze on the ship was forbidden yet obtainable. There was always someone who had a stash for nights when a little 'medicinal nourishment' was needed. The point was never to overindulge. To get caught with alcohol meant extra duties, less flying and being placed 'in hack'. This meant that you were to remain aboard the ship during the next port call. After months at sea, the last thing you wanted was to be left on the boat while your squadron mates were on the beach. There was one time when alcohol was allowed, in fact, distributed on the ship. Beer Day came about when the ship had gone more than forty-five days at sea without a port visit. We had spent more than ninety days, and were rewarded with two beers a piece,

distributed under the supervision of the military police, and it had to be consumed on deck."
– Frank Furbish, former US Navy fighter pilot

"I remember J.T. 'Boats' Boatwright of the V-5 Division [USS *Enterprise* (CV-6)] from Thanksgiving Day in 1942. We were in Noumea, New Caledonia, being repaired by the Seabees after being hit pretty hard in the Battle of Santa Cruz. Our planes were on the beach and 'Boats' and I were sent ashore to repair the synchronizer on one of our SBDs, with orders to stay on the job until the plane was operational. I was 3rd class having just come aboard in Pearl the first part of October. 'Boats' was the old-timer at 1st class.

"We finished the job and went to an Army mess hall. The Navy didn't have much there at the time. Their cook said they had cleaned him out and he didn't have any food left. After 'Boats' told him our plight he said 'wait a minute,' and came back with two bottles of beer and two oranges. We went outside and sat under a coconut tree and ate our Thanksgiving dinner. 'Boats' said: 'You can tell your grandkids about this in years to come.' It's the only Thanksgiving dinner I remember."
– Wilbur West, formerly with V-5 Division, USS *Enterprise* (CV-6)

"Beans, beans and more beans. And sometimes potatoes for breakfast. But Navy food was seldom disappointing to me. Meals have never been high on my list of priorities. As a 'near vegetarian', I was always popular as a chow buddy, being quite willing to swap my steaks, ham or pork chops, for desserts. Navy breads and pastries were usually tasty, and even the reconstituted milk wasn't bad if it was cold enough. Holiday meals were truly exceptional, with all the trimmings except flowers on the tables. The 'chemist coffee' ensured my life-long aversion to that beverage, and I doubt that anyone ever really adjusted to standing in line so long for meals. Even today I avoid eateries with slow service.

Still, many of us carried books to read and others studied for exams, so the time in the lines was not entirely wasted.

"It wasn't all work, by any means. Movies were often shown on the hangar deck, and impromptu shows were staged by talented personnel. The forward deck elevator, lowered about half way, made a satisfactory stage, and the ship's sixteen-piece band was exceptionally versatile. There was a glee club, a hillbilly band, a ship's radio station with disc jockeys, and a well-stocked library. There were Judo classes, basketball, boxing and weight-lifting . . . even model airplane building.

"After extended periods 'on station' conducting missions off the coast of Korea, we would enter the harbor at Yokosuka, Japan for replenishment of supplies, to off-load battle-damaged aircraft, and to spend a few precious days on liberty. The first visit there was a culture shock, as we were not prepared for the food, the language, or even the procedure for buying souvenirs. We quickly adapted and, although we seldom had much spending money, prices there were very low at the time, and we found many items that were useful for shipboard living. There were inexpensive shower shoes which were held on only by rubber strips that fit between the first two toes. They felt strange at first, and if one tried to back up, they would slide off, and they soon became known as 'go-aheads'. In addition to shopping, we took side trips on the efficient railroads to places such as Tokyo and Kamakura. During one visit to Yokosuka, a team of Japanese carpenters came aboard to repair flight deck damage caused by a prop plane that had crashed on deck. Our deck had a layer of teak wood on top, chosen because when hit, it would harmlessly 'powder' rather than fragment into dangerous splinters. The ship carried spare teak planks, and it was fascinating to watch these skillful workmen in action and the strange (to us) tools they used.

"The *Kearsarge* ported in Hong Kong, and we were immediately surrounded by small craft with vendors eagerly shouting up to us about the wares they were selling. We had been alerted about the poverty prevailing, however, seeing at first hand the contrast between the high-rise buildings, upscale stores, well-dressed and prosperous people . . . and the wretched people crammed aboard the tiny boats, cramped together in the scummy waterways, was truly depressing. Sailors seldom have much spending money, but we were rich in comparison to these boat people, and it was hard not to feel guilty about our good fortune in having been born Americans.

"A legendary Hong Kong business lady named Mary Soo had organized a group of industrious Chinese to serve as side-cleaners for the US Navy ships. This was among the most miserable chores involved in maintaining an aircraft carrier, yet here were people who were delighted to have work of any sort, eagerly scrubbing away while we were off exploring the wonders of Hong Kong.

"Among our discoveries were store-fronts that were obviously aimed at attracting English-speaking tourists. One such was a shoe shop with the sign: NO-SQUEAK JOHNSON. Other stores had tailors who could make low-cost suits, virtually overnight. There were a lot of wood-carvers too, so some of the crew returned to the ship with garments and large carved dragons, bragging about their bargains. Imagine their chagrin later, when the sleeves came off their hurriedly-made suits, and cracks began to appear in the carved dragons after a few weeks exposure to the sea air.

"The Navy has a policy of rotating work assignments among the men. The reasoning was that there would be enough overlap in qualifications so that all requirements could still be met if part of the crew was on leave ashore. It was inevitable that more than a few people were given jobs for which they were entirely unsuited. Case in point: Algiers had an unsavoury reputation, and it surprises me still that we were

SHRIMP SCAMPI
(yield: 100 portions) 4 pans

Ingredients:
25 lb shrimp, raw, peeled, deveined, thawed
6 1/2 cups (3 lb 4 oz) butter or margarine
3 1/3 tbsp (1 1/4 oz) garlic, dehydrated
1 cup (1/2 oz) parsley
3 tbsp (2 oz) salt
1 tbsp pepper, black
2 tbsp lemon juice
8 3/4 cups (3 lbs) tomatoes, fresh, chopped
2 qt (2 lbs) bread crumbs, dry

Method:
1. Rinse shrimp; drain. Place 6 lb 4 oz shrimp in each pan.
2. Combine butter or margarine, garlic, parsley, salt, pepper and lemon juice; simmer 1 minute.
3. Add tomatoes to butter mixture; stir lightly but thoroughly. Pour 3 1/4 cups tomato-butter mixture over each pan, thoroughly coating shrimp.
4. Sprinkle 2 cups bread crumbs evenly over top of each pan.
5. Bake 30 minutes at 400°F. Serve immediately.

All those forward, move aft. All those aft, move forward. All those amidships, mill around and create confusion.

HMS *Illustrious* General Mess menu for dinner, Monday, 7 June 1999

Baked Gammon with
Cumberland Sauce
Chicken Maryland with
Sweetcorn Fritters
Lamb Madras with Plain Rice
Swiss Cottage Pie
Savoury Vegetable Stew
Roast & Boiled Potatoes
Straw Carrots, Brussel Sprouts,
Creamed Swede
Fruit Pie with Custard Sauce
Selection of Cold Sweets

Below: Sailors from the USS *Randolph* (CVA-15) paying their respects to the ladies of this French Riviera beach in the 1950s.

even allowed to go there. However, we were lectured in advance of our visit about cultural differences and warned about potential concerns, which ranged from unsafe drinking water, to being served dog meat, to encountering pickpockets. In its wisdom, the Navy decided to send me to the Casbah, of all places, on Shore Patrol duty. I had no such qualifications, but attended a forty-five-minute crash course on becoming an instant policeman, in which it was stressed that we must, at all costs, avoid agitating the natives, and we must not even look at the veiled women. We were told that our main purpose was to keep our sailors, especially the heavy drinkers, in line and out of trouble. I was already apprehensive when they issued us Shore Patrol arm bands and billy clubs, explaining, 'We can't give you sidearms because the natives might steal them and use them against you.' Off I was sent, in company with another unqualified sailor, to take the midnight shift. 'Don't worry,' we were told. 'We'll send a resident Casbah policeman along with you in case there might be any language problems.' It turned out that the fellow spoke several tongues, none of them English. We made the rounds of several hot spots and, fortunately, encountered no drunks. We had decided that if we came upon any fights, we would simply let them kill each other. Mercifully, the hours passed uneventfully, and we returned to the ship, grateful to be back aboard.

"Shopping was fun almost everywhere and, except for the French Riviera, we could generally afford at least a few souvenirs. Compensating for the sky-high prices on the Riviera were the spectacular beaches and the even more spectacular swimming suits. Bikinis were a recent phenomenon that we had seen only in magazines back home, but in France they were everywhere. As a happily-married man, I was not shopping for female companionship, but must admit it was intriguing

214

seeing those totally uninhibited *demoiselles*, some of them topless."
– Bill Hannan, former US Navy jet engine mechanic

"A couple of months after the war [World War II] started, the *Enterprise* paymaster asked for the confidential advice of the senior squadron officers. He had noticed that seven enlisted men on the 'Big E' were sending allotments to their wives, all to the same address. It was, however, apparent that only one woman lived at that address. Should the men be told or not? The final consensus was that the matter should be kept secret and the men should not be told as they would be devastated and their essential performance would suffer."
– Jack "Dusty" Kleiss, former US Navy pilot

"Imagine being part of 6229 men having Thanksgiving together. Now imagine trying to shop for all the food needed to feed these hungry men, not only for Thanksgiving, but for an additional thirteen days.

"This group was made up of 4710 soldiers who were headed home after two or more years of fighting the war in Europe, and 1519 sailors who made up a skeleton crew that was necessary to sail the aircraft carrier *Enterprise* (CV-6) that was transporting them home.

"The *Enterprise* had participated in and survived twenty-three Pacific battles in World War II. She was reported to have been sunk six times by the Japanese, and had seen more action in the Pacific than any other ship. Operation 'Magic Carpet' brought her home to a new assignment in the Atlantic, to bring troops home from Europe.

"I was honoured to be part of the crew as we sailed from Pearl Harbor, through the Panama Canal and north along the Atlantic coast to New York. Along the way the *Enterprise* had her last 'flight quarters' and her planes flew off to land bases.

"Following a great welcome in New York

harbour, called Operation Broadway, and the honouring of the fleet, *Enterprise* set sail for Boston to be fitted out. The hangar deck had 700 folding bunks five tiers high installed to accommodate 3500 men. The remaining men were housed below decks. Next came the challenge of feeding our new 'cargo' as well as the crew.

"What and how much food would be needed? I managed to get a copy of the 'shopping list'. A special Thanksgiving menu was planned because we would be in Southampton, England on that day with our returning soldiers on board. These 4710 troops were loaded on and assigned to quarters in 48 hours. The 500 tons of food loaded on in Boston consisted of: 14,000 dozen eggs, 30 tons cold storage beef, 5 tons fresh pork, 6 tons butter, 3 tons turkey, 4 tons ice cream mix, 36 tons potatoes, 3 tons tomatoes, 4 tons cabbage, 15 tons fresh fruit, 10 tons coffee, 25 tons sugar, and 50 tons flour. We distilled 75,000 gallons of water each day, with 15,000 going into the boilers.

"We arrived in New York on 1 December after a reasonable crossing. Our passengers were quickly unloaded and new supplies were loaded to fill our empty pantry in preparation for our next crossing scheduled for 8 December. The trip over was uneventful and, again, nearly 5000 soldiers were loaded aboard. The plan was to be home a few days before Christmas.

"The Atlantic had other plans for us. We ran into a storm, supposedly the worst in sixty years. Waves crashed over the length of the flight deck, 830 feet long and 60 feet above the water. Life rafts were torn loose, an anchor was ripped off, and water sprayed in on the now seasick soldiers (and a few sailors too). We were actually blown back towards England, and some of the *Enterprise*'s old war scars were reopened.

"We did make it home for Christmas. We sailed past Staten Island at 5.30 on Christmas Eve 1945."
– John R. Dunbar, formerly assigned to the USS *Enterprise* (CV-6)

Americans can eat garbage, provided you sprinkle it liberally with ketchup, mustard, chili sauce, tobasco sauce, cayenne pepper, or any other condiment which destroys the original flavor of the dish.
– from *Remember to Remember*
by Henry Miller

Life, within doors, has few pleasenter prospects than a neatly arranged and well-provisioned breakfast-table.
– from *The House of the Seven Gables*
by Nathaniel Hawthorne

I love the Flora-Bama. Some people say it's a dive, but I think it's great, especially in the winter when it's just the locals and the flight students. It's right down there on the beach. Completely tawdry; I love it.
– Shannon Callahan, Naval Flight Officer student, NAS Pensacola

Recipes courtesy the USS *John C. Stennis*, 1999

Menus courtesy HMS *Illustrious*

AIR STRIKE THE FALKLANDS

Good judgement comes from experience. Unfortunately, experience usually comes from bad judgement.

In the pathway of the sun,
In the footsteps of the breeze,
Where the world and sky are
one, / He shall ride the silver
seas, / He shall cut the
glittering wave. / I shall sit at
home, and rock; / Rise, to heed
a neighbour's knock; / Brew my
tea, and snip my thread;
Bleach the linen for my bed.
They will call him brave.
– *Penelope*
by Dorothy Parker

'And, Sir, the secret of his
victories?' / 'By his
unServicelike, familiar ways,
Sir, / He made the whole Fleet
love him, damn his eyes!'
– from *1805*
by Robert Graves

Right: A survivor of the Argentine attack on the Royal Navy destroyer *Sheffield* being rushed from a rescuing helicopter to the sick bay of the carrier HMS *Hermes* during the Falklands War.

ON 2 APRIL 1982, when the British government directed that a Royal Navy task force be sent to the South Atlantic to recapture the Falkland Islands from the Argentinian occupation forces of General Leopoldo Galtieri, world opinion was generally opposed. In Britain the more interesting debate for many was military rather than political. How could a relative handful of Sea Harriers, a weapons system not without its detractors at the time, stand up to a force of nearly 200 Argentinian Mirages, Skyhawks, Etendards and more, in a part of the world where operating conditions, never mind the odds, were truly terrible?

Twenty years before the Falklands campaign, Canadian-born Nigel David "Sharkey" Ward was an Officer Cadet at the Britannia Royal Naval College, Dartmouth. In his naval career he went on to become a qualified Air Warfare Instructor. Ward trained on and flew Hunters, Sea Vixens and F-4K Phantoms. In 1981 he became commander of 801 Naval Air Squadron in HMS *Invincible*. In July of that year he was given the opportunity, and the challenge, of proving the capability of the Sea Harrier and its radar and weapons system under operational conditions. He and his squadron had to find out just how good the airplane was and how suitable it was for its job. Ward's years of experience with the Sea Harrier, his determined efforts to get the airplane operational, and his part in the fight to keep the fast jet Fleet Air Arm alive and airborne, earned him the title "Mr Sea Harrier". In extracts from his book, *Sea Harrier Over The Falklands*, Commander Ward offers a fascinating look at the air campaign of the 1982 conflict.

"The 801 Squadron recommendations for high-level bombing against Port Stanley airfield had been accepted somewhat reluctantly by the Flag. This reluctance was not really surprising, bearing in mind that the attack relied completely on the capabilities of the [Blue Fox] radar and the Navhars [Sea Harrier Navigation Heading Attitude Reference System] for its success — bearing in mind, too, what the Flag thought of such things.

"In reply to our recommendations and after due deliberation, SAVO [Staff Aviation Officer: a member of an Admiral's staff specially appointed to advise the Admiral on all operational and training aviation matters] put forward several fully anticipated questions: how were the pilots going to ensure that they missed the town with their bombs? Would the line of attack take them over the town? And so on. Finally, realizing that my team had no intention whatsoever of risking the lives of the Falkland Island community, and after having been briefed in detail about the profile to be flown by attacking aircraft, the Staff agreed to the proposal. But 801, having generated and recommended the new delivery, were not the first to be tasked with trying it out. That didn't matter very much to me. The main thing was that now we could do our bit to keep the enemy awake and on the alert; the less rest the Argentines ashore got before our amphibious troops landed, the better.

"To that end — denying the enemy some sleep — we decided in *Invincible* to set up a mission that would give the opposition more than a little to think about in the middle of the night. There were two major detachments and airstrips away from Stanley. On East Falkland there was Goose Green, close to the settlement of Darwin. On West Falkland there was Fox Bay. With the Task Group well to the east, the distance to Fox Bay was about 225 nautical miles as the crow flies. The plan was very simple. I was to get airborne on my own in the small hours of the night and deliver a Lepus flare [a six million-candlepower flare, dropped or tossed from fast jets to illuminate targets for reconnaissance or night-attack purposes] attack against each defended airstrip. It was hoped that this would give the troops on the ground the false

'Good show!' he said, leaned his head back and laughed. 'They're wizard types!' he said, and held his beer Steadily, looked at it and gulped it down / Out of its jam-jar, took a cigarette / And blew a neat smoke ring into the air. / 'After this morning's prang I've got the twitch; / I thought I'd had it in that teased-out kite.' / His eyes were blue, and older than his face, His single stripe had known a lonely war / But all his talk and movements showed his age / His whole life was the air and his machine, / He had no thought but of the latest 'mod', / His jargon was of aircraft or of beer. / 'And what will you do afterwards?' I said, Then saw his puzzled face, and caught my breath. There was no afterwards for him, but death.
– *Fleet Fighter*
by Olivia FitzRoy

impression that a Task Force landing was under way — the flares apparently providing illumination prior to an assault.

"The mission was very straightforward. All I needed to do was enter the attack co-ordinates for each target into my navigation computer, follow the resultant HUD [Head-Up Display] directions to the target areas, check the accuracy of the navigation using my radar, and drop the flares. Outbound to the targets I would fly direct to Goose Green from the climb-out point which would take me close to Stanley. On the return leg I would approach the Carrier Group from 240°, that is, well clear of the islands, let down to low level and then run in to *Invincible* for recovery on board.

"JJ [Captain J.J. Black, skipper of *Invincible*] was pleased with the idea of the planned mission: 'Anything we have missed, Sharkey? You know that the weather isn't too brilliant out there?' The weather round the Task Group was certainly not very sharp, with high winds, towering cumulo-nimbus and snow showers. But over the islands the forecast was for much fairer conditions. The mission would therefore present no problems and it hurt my pride even to think that I might have a problem getting back on board. I shut my mind to snow showers and thought only of dropping the flares.

" 'No, thank you, Sir. I'm very happy with all that. Can we just make sure that the Flag lets all our surface units know exactly what's going on? One lone aircraft approaching the force at night from the south-west is just asking for a 'blue on blue'. ('Blue on blue' means own forces shooting down own aircraft.)

" 'Don't worry, we'll ensure that they let all units know.'

"There was practically no self-briefing required for the mission. It was a question of kicking the tyres, lighting the fires and getting airborne.

"I arrested my rate of climb after launch at 500 feet, settled the aircraft down to a height of 200

feet above the waves, set 420 knots and departed the ship on a north-westerly heading for 70 miles. As my eyes became fully attuned to the darkness the sea surface became visible, with the healthy wind kicking up a lot of spray and phosphorescence and clouds towering around and ahead of me like huge cliffs. Inside the cockpit I had dimmed the lighting almost to extinction, including the geometrical patterns on the HUD. Arming missiles and guns, just in case, I reached the climb-out point and headed for the stars, turning to port as I did so to get on track for the first destination.

"The aircraft climbed at a cool 30° nose up for the first 15,000 feet and very quickly I was throttling back at Mach 0.85 and 35,000 feet, enjoying the view. Under the black, star-speckled sky and towards East Falkland, the cloud thinned and dissipated. Although it was the middle of the night, Stanley was well-lit and could easily be seen from the top of the climb.

"Dick Goodenough [801 Squadron Air Engineer Officer] had given me the jet with the most sought-after radar and Navhars — 004. The engineers kept full records of the performance, accuracy and reliability of each radar and Navhars system. This attention to detail, which included comprehensive post-flight briefs, paid excellent dividends. But although all the aircraft performed well, 004 seemed to be the pick of the bunch. The avionics were functioning perfectly. Before passing just to the south of Stanley, I checked the navigation system accuracy with my radar and, until commencing descent for Goose Green, used the radar in its air-to-air mode to check the skies to the west for enemy fighters. There were none, nor did I expect any. I was enjoying being alone and thought of my lads back at home — they would have given anything to have been on this flight.

"I chose to run into Goose Green up Choiseul Sound and had an excellent surface radar picture to confirm the accuracy of my desired flare-drop

position. Having checked the Lepus selector switches and pylon, I monitored the Navhars readout, which was telling me time-to-go to release, and the HUD, which was guiding me unerringly to the target. Pickling the bomb button, I felt the flare body leave the aircraft, applied power and commenced the second climb of the night to high level.

"After about 20 seconds I rolled the Sea Jet inverted in order to see the flare ignite. Goose Green lies on a narrow isthmus of land which separates Choiseul Sound from Grantham Sound. When the flare went off it seemed to be in the right place — right on target. Now for Fox Bay.

"The second target was well-covered in thick cloud, but that made no difference to the attack profile. My radar confirmed what the Navhars was telling me and I had no doubt at all when I pickled the bomb-release button that the Lepus was again bang on target. It was as easy as riding a bike. Having seen the glow from the flare through the cloud, I turned to the east in the climb and set off on the long dog-leg home.

"At 140 miles to go to the deck, I was suddenly illuminated with fire-control radar from a ship below. The relative silence of the cockpit was shattered by the radar warning receiver alarm. I had detected a vessel on radar earlier and had presumed it was on detachment from the Task Group.

"Thoughts of the Argentine Navy firing Sea Dart at me flashed through my mind. I broke hard to starboard and descended to the south-west away from the contact which was only 20 miles to the north. But I was still locked-up by the fire-control radar, and having taken initial avoiding action I started to analyze the threat. It certainly wasn't Sea Dart — the noise in my earphones didn't have the right characteristics — and so I was pretty safe now, 30 miles south. I turned port onto east, arresting my descent as I did so, and climbed back up to 30,000 feet. I would circumnavigate the ship that had shown such an unhealthy interest in me.

At the same time I would interrogate it on radio.

"I found the culprit on the pre-briefed Task Group surface/air frequency. 'Warship illuminating lone Sea Harrier, come in.'

"The warship had obviously not been briefed on the SHAR [Sea Harrier] mission and was unwilling to believe I was who I said I was. Being now a little short of fuel after three-and-a-half climbs to high level, I left my personal signature encoded in terms of four-letter words and bid them goodnight. If I had been on a slightly different track and hadn't had my radar information to tell me where to evade to, I could have been engaged by my own forces. So much for the promise to pass information to all ships.

"By the time I began my final descent towards the Carrier Group I was back amongst the clouds. They were massive and very turbulent. After I had descended to low level and was running in to the expected position of the ship via the safety lane, I called, '004, on the way in. Estimating 280°, 25 miles. Over.'

"Tony [Direction Officer in *Invincible*] was immediately on the air.

" 'Roger, 004. Read you loud and clear. I have no contact on you, repeat no contact. Clutter from snow clouds too intense.' He was concerned. Good old Tony; there's a man you can really trust. He'd do anything to get his pilots down safely.

" 'Roger. I'll conduct my own approach and call out my ranges to go.' I was feeling confident thanks to two important facts. Firstly, when *Invincible* gave a ship's estimated position for the recovery of aircraft, you could bet your pension on her being in that position when you returned from your flight; especially in bad weather. So I was very sure in my mind that I could find the deck using my Navhars information. The second fact was that I had practised self-homing to the deck on many occasions, and we had also carried out the trials on the software for self-homing when ashore in the Trials Unit. It was no higher

work load for the pilot than following instructions from the ship's precision approach controller.

"On my radar screen, the *Invincible* 'position destination marker' that I had selected on my nav computer sat less than 2 miles from one of the ship contacts in view. I had already programmed the 'marker' with the ship's pre-briefed recovery course and speed and was happy to see it was holding good formation on the contact nearest to it. That had got to be *Invincible* — I hadn't enough fuel left to make mistakes. There was enough for one approach only.

"It was a simple matter to update the radar 'marker's' position by fixing the radar onto the contact. The 'Self-Controlled Approach' programme in the Navhars computer software was provided so that pilots could safely carry out their own precision approach to a chosen destination. My chosen destination was the ship, and as I lined up 5 miles astern of what I thought was *Invincible*, I selected the precision approach mode on the HUD. I also locked the radar onto the ship to keep the 'destination' information as accurate as possible. Now I was all prepared for recovery to the ship.

" 'Five miles on the approach', I called.

" 'Roger, still no contact.' Tony must be sweating buckets down there.

"I was at 800 feet and the world outside was black. Approaching 3 miles I prepared to commence descent. The radar was firmly locked on to the contact ahead.

" 'Three miles.'

" 'Still no contact.'

"Was I on the right ship? I began to wonder as I started down the slope. My jet was being tossed around a bit by turbulence from heavy clouds, which would certainly account for the clutter Tony had mentioned. There was no other course but to wait and see. I cleared the doubts from my mind.

" 'Have you now at 1 1/2 miles. On the glide slope.' Tony sounded relieved. I was relieved.

In the 1982 fight for the Falkland Islands, the Argentine Navy had its Escuadrilla Antisubmarina equipped with six Grumman S-2E Tracker aircraft that had originally been supplied by the United States for anti-submarine duty, and four of them were operated from the carrier *25 de Mayo*. As a part of Task Force 20, it was their job in Operacion Rosario to maintain a continuous airborne watch and seek out any British submarines or surface ships whose presence might interfere with the Argentine operation. In the operational period against the British, the squadron flew 112 sorties, making contact with British warships approximately forty times, all to little effect.

Left: In the 1982 Falklands War, Commander Nigel David "Sharkey" Ward, DSC, AFC, RN commanded 801 Naval Air Squadron in HMS *Invincible*. He had been senior Sea Harrier advisor to the Command, and in the conflict he flew more than sixty missions.

3 Escuadrilla de Caza Y Ataque could muster only eight fully-operational Douglas A-4Q Skyhawks when required to embark on the Argentine carrier *Veinticinco de Mayo* in March 1982. The squadron had lost six of the planes to accidents; three of them during earlier deployments on the carrier. As part of Task Force 20, the carrier was involved in the protection of Argentine assault forces in their efforts to retake the Falklands on 2 April. As it happened, the Skyhawks were not needed in the landings and were returned to their land base when the islands had been secured. The situation altered significantly, however, by the middle of the month, and the planes were once again embarked on the carrier. A British task force was underway to the area and the Skyhawk squadron looked forward to the possibility of air and sea battles with the new enemy.

On 2 May, with the British task force just 150 miles from the Argentine carrier, all eight Skyhawks were loaded with iron bombs and their pilots briefed for an attack against the British fleet. The attack was not launched. It may be that very light winds of that morning prevented the heavily-laden A-4s taking off at maximum weight. Another view, however, is that the searching Argentine S-2E Tracker aircraft simply lost its earlier radar contact with the Royal Navy carrier *Invincible* and was unable to reacquire contact with it. Later that day the Argentine cruiser *General Belgrano* was torpedoed by a British submarine and was lost. The Argentine carrier was quickly withdrawn to home waters, and the great sea battle its crew and airmen had anticipated did not happen.

"Tony continued with his calls all the way to half-a-mile. He had passed the wind over the deck as 40 knots gusting 50. It felt like it in the cockpit, too. The buffeting increased as I got lower.

" 'Half-a-mile.' My head-up information said the same. I delayed selecting hover stop for a few seconds because of the strong head-wind, then nozzles down, power going on. At a quarter of a mile I called 'Lights'. And there, behind the radar cross in the HUD, appeared the ship's island. As usual the cross was just about on Flyco. Radar off and concentrate on controlling the jet. As I was moving sideways over the deck from alongside, the wind backed through 30°. I ruddered the nose into it before setting onto the deck with an uncharacteristic thud.

" 'That's my excitement over for the night', I thought. It was 0400 hours, and a long day lay ahead."

"By the time June arrived we had grown well-used to operating in the unpredictable South Atlantic weather, and to our routine air transits to and from the islands for CAP [Combat Air Patrol]. Combat opportunities against the Argentine Air Force and Navy pilots were becoming less and less frequent. But there was never a moment to relax vigilance on CAP because the enemy had not given up. On the contrary, they still held all the strategic points on the islands, and although Jeremy Moore's [Land Forces Deputy] men had landed they were not yet in a position to displace the occupation forces.

"Intelligence gleaned from various sources indicated that Port Stanley was still being supplied by air on a nightly basis by AAF Hercules aircraft, and the same aircraft were making regular use of the landing-strip at Fox Bay to support their logistic re-supply efforts. Without a radar look-down-over-land capability, there was little that we could do to intercept these night resupply flights, especially with the Task Group positioned

between 200 and 230 nautical miles to the east of San Carlos Water. The majority of the Sea Harrier effort by day continued to be spent on maintaining CAP stations around the beach-head and, at that range from the carriers, our on-task time had come down to about 25 minutes over the Sound (and that was pushing it). It was, nevertheless, effort well spent, and the SHAR's continuous presence to the west of the Amphibious Operating Area continued to deter and disrupt enemy air attacks.

"On 1 June my section had launched after daybreak and we had carried out our CAP duty over Falkland Sound without any sign of trade. Everything appeared quiet in the skies to the east as our two jets commenced the climb—out to the north of San Carlos en route to the ship. Our home transit would be at 35,000 feet and, in the climb, Steve [Thomas] kept his Sea Jet in neat battle formation on my starboard beam.

"HMS *Minerva* was the Local Area Control Ship and we had not yet switched from her radio frequency. The craggy north coast of East Falkland Island was well below us when my helmet headphones crackled as *Minerva* called us up.

" 'I just had a "pop-up" contact to the north-west of you at 40 miles. Only had three sweeps on the contact and it has now disappeared. Do you wish to investigate? Over.' I could tell from the controller's voice that he really felt he had something but didn't want us to waste our time if he was wrong. After all, 40 nautical miles was a long way in the wrong direction when our aircraft could be running low on fuel.

"He needn't have worried; neither I nor Steve would dream of turning down the slimmest chance of engaging the enemy. Before he had finished his call I had already started a hard turn to port with Steve following in my wake. I was flying aircraft side number 006 and its radar was on top line.

"In the turn and using the radar hand-controller thumb-wheel, I wound the radar antenna down to just below the horizon, and as we steadied on a north-westerly heading there was the target. The green radar blip stood out as bold as brass in the centre of my screen at just less than 40 miles.

" 'Judy! Contact at 38 miles. Investigating.' This was a chance we couldn't miss.

"I decided to try to lock the radar to the target to get some accurate height information on it. The radar locked easily, telling me as it did so that the target had to be a large one at that range, and gave me what I wanted to know, a height difference of 4000 feet. I was at 12,000; that made the target at 8000. I broke the lock so as not to alert the target (he would be listening out on his radar warning receiver) and wound our speed up to 500 knots in a shallow descent. Target range decreased rapidly to 34 miles, 30 miles — then it seemed to hold. There was only one possible reason for that.

" 'Steve, I think he's turning away. He's now at right 10° at 28 miles. 4000 below.'

" 'Roger. Contact.' Good! Steve also had contact.

"I locked the radar again. Still about 4000 feet below, but we had already come down to 10,000 in descent.

" 'He's definitely turned away and he's descending. Must have seen us.' The shore control radars of the Argentine forces must have monitored the start of our intercept and then passed the information to the target.

"It was now a race against time and fuel. *Invincible* was over 200 miles away and we should have been heading home. But there was an easy alternative. I called *Minerva* on the radio.

" 'We may be too short of fuel to get back to Mother after this. Can you ask the assault ships to prepare to take us on board in San Carlos?'

"A short pause before *Minerva* came back. 'We have decks ready to take you if you need it.'

" 'Roger. Please check that weapons will be "tight" in the missile zone if we pay you a visit.' I knew it wasn't necessary to remind *Minerva* of that, but it was better to be safe than sorry.

" 'Roger. No problem.' The ship-borne controller

then settled down to monitor the chase — if we destroyed the target it would all be as a result of his sharp radar pick-up and concentration.

"Having sorted the fuel problem out in my mind, I could give my full attention to tracking the target, which was now heading west and had descended to low level below the cloud. We were still cracking along above the cloud in brilliant sunshine. I checked my missile and gun switches; safety flaps were up and everything was live and ready to go.

"The sea state was markedly rough and as we continued to close the fleeing target from above, I wished that the Flag and all his Staff could have been in the cockpit with me to witness the radar's performance in these look-down conditions. The radar was holding contact with the target on every sweep.

"We were now approaching the cloud-tops at 6000 feet and catching the fleeing aircraft fast. To stand any chance at all of survival, the slower target ahead would need to stay in the cloud layer and try to evade us with hard manoeuvring. But there was little chance of that being successful either, since I had had a lot of practice against large evading targets in cloud by day and night.

" 'Steve, you'd better stay above the cloud until I am visual with the target below.' If he popped up through the cloud layer, Steve would get him.

"Roger. Still good contact at 9 miles.'

"I descended steeply into the cloud layer, breaking through the bottoms at 1800 feet. I still had the enemy on my radar screen; a fat blip at 6 miles and closing fast. I looked up from the radar and flight instruments and there it was at 20° left, a Hercules heading for the mainland as fast as it could go. It was at a height of about 300 feet above the waves.

" 'Tally ho, one Herky-bird! Come and join me down here, Steve.'

"I closed the four-engined transport fairly quickly and when I felt I was just within missile range and had a good growl from the seeker-head I fired my first Sidewinder. As usual, it seemed an eternity before it came off the rails and sped towards its target. I had locked the missile to the left-hand pair of engines. The thick white smoke trail terminated after motor burn-out and the missile continued to track towards the Hercules' left wing. I was sure it was going to get there, but at the last minute it hung tantalisingly short and low on its target and fell away, proximity-fusing on the sea surface below.

"There was no mistake with the second missile. I locked up the Sidewinder on the target's starboard engines, listened to its growling acquisition tone, and fired from well under 1 1/4 miles. It left the rails with its characteristic muffled roar and tracked inevitably towards the right wing of the Hercules, impacting between the engines. Immediately, both engines and the wing surface between them burst into flames.

"Our fuel state was now getting marginal, to say the least. The job had to be finished, and quickly. Otherwise the Argentine aircraft might still limp home and escape. I knew the Hercules had an excellent fire suppression system in the wings and we couldn't let it escape now.

"I still had more than 100 knot's overtake as I closed to guns range and pulled the trigger. My hot-line aiming point was the rear door and tailplane and all of the 240 rounds of 30mm high-explosive ammunition hit their mark. There were no splashes in the sea below.

"As I finished firing, and with its elevator and rudder controls shot away, the large transport aircraft banked gracefully to the right and nose-dived into the sea. There could have been no survivors.

"Pulling off the target hard to port, I called *Minerva*.

" 'Splash one Hercules! Well done on spotting it!'

"As the controller's excited voice came back on the air, I could hear the cheers of the Ops Room staff in the background. They also knew that the

Every take-off is optional. Every landing is mandatory.

It's always better to be down here wishing you were up there, than up there wishing you were down here.

224

Hercules force had been running supplies into Stanley on a daily basis, usually at night and always at very low level. To our ground forces they were a high-priority target. This Hercules had shown some complacency by popping up to 8000 feet, and had paid the price.

" 'Nice one.' It was the *Minerva* controller. 'Do you wish to land in San Carlos? We'd all like to see you both.'

" 'Roger. Wait. Steve, check fuel.'

" '2100,' came the reply.

"I had a couple of hundred less and the ship was 230 miles away. We couldn't rearm my aircraft on board the assault ships and as I thought we could just squeeze home to *Invincible*, I decided to turn down *Minerva*'s kind invitation.

" 'Sorry, we can just make it back to Mother, so we'd better do that. Thanks anyway for standing by.'

"We were already in the climb and in less than half an hour were touching down on deck. There had been a tail-wind so we ended up with about 400 pounds of gas in the tanks — more than I had expected.

"During the preceding weeks there had been much talk in our crewroom about knocking down a Hercules or similar large transport. Given the chance, everyone favoured flying up alongside the cockpit and signalling to the crew to jump out. We felt no animosity towards the Argentine pilots; they were just doing what they had to do, and if their lives could be spared then they would be. Sadly there had been no time for such chivalry on this occasion. The choice was chivalry, thus possibly giving the enemy aircraft the opportunity to survive and maybe running our Sea Jets out of fuel, or a quick kill. Circumstances and, in particular, fuel states dictated that it had to be the latter choice. I didn't lose any sleep over it, but wished that we had had more time to play with.

"The Hercules intercept turned out to be the last kill of the air war in which the Sea Harrier weapon system was able to play its full part."

Below: In late May 1982 a Sea Harrier is being prepared aboard HMS *Hermes* for a sortie in the Falklands conflict.

IT WAS DURING THE KOREAN WAR that helicopters were first employed on a large scale in an aggressive vertical envelopment concept. The lifting and placing of combat units where they needed to be, rescuing downed air crews, and quickly evacuating wounded to rear area medical care, were tried and proven in that conflict of the early 1950s. There were, however, no dedicated vessels for the support of the rotary aircraft in these important functions, thus fully integrating them into US amphibious operations was only a theoretical possibility. The theory moved closer to reality later in the '50s when, in an interim move, some elderly *Essex*-class aircraft carriers that were being phased out of fixed-wing aircraft operations, were redesignated as LPHs (Landing Platform Helicopters). In 1961 the first purpose-built ship for this role, the USS *Iwo Jima* (LPH-2), entered service. Designed with economy in mind, the *Iwo Jima*-class ships lacked a lot of the more refined features of the large fleet aircraft carriers. In performance, however, they could operate at a sustained speed of 20 knots. Like the current *Invincible*-class carriers of the Royal Navy, the flight decks were not fitted with catapults or arresting equipment, operating mainly to support ASW helicopters and AV-8 Marine Corps Harriers.

The *Tarawa*-class LHA (Landing Helicopter Amphibious) carriers, such as the *Nassau* (LHA-4), are larger than the LPH ships and utilize aircraft like the CH-46E Sea Knight helicopter to land Marine assault troops in all weather conditions, as well as transport their equipment and supplies. Completed in the 1970s, the ships of the *Tarawa* class include the namesake (LHA-1), and *Saipan* (LHA-2), *Belleau Wood* (LHA-3), *Nassau* (LHA-4) and *Peleliu* (LHA-5). In addition to the Sea Knights, LHA ships carry CH-53 Sea Stallion helicopters, AH-1 Seacobra gunship helicopters (which were developed from the Huey line of Vietnam–era gunship choppers) and UH-1 utility helicopters. They can also support the CH-53E

Super Stallion helicopter. Some LHAs can support the AV-8B Harriers operated by the Marines.

The modern assault ship of the US Navy at the start of the twenty-first century is called a Primary Landing Ship. It looks a lot like a small aircraft carrier, and its main role is to bring troops to hostile shores. Its complement of US Marine expeditionary units is then brought to the beach by a combination of helicopters, conventional landing craft, and LCACs (Landing Craft Air Cushion). These are the vessels of the *Wasp* class and include *Wasp* (LHD-1), *Essex* (LHD-2), *Kearsarge* (LHD-3), *Boxer* (LHD-4), *Bataan* (LHD-5) and *Bon Homme Richard* (LHD-6). In an expanded role, these impressive assault ships serve to project US power and sea control. To that end they employ anti-submarine helicopters like the Kaman SH-2G Seasprite, which also has an anti-surface threat capability, including over-the-horizon targeting. Additionally, the assault ship carries the highly capable Boeing McDonnell Douglas AV-8B Harrier jet VSTOL ground attack aircraft. Powered by a Rolls-Royce vectored-thrust turbofan, the AV-8B is a key weapons system of the US Marine Corps.

The Seasprite helicopter increases and extends the assault ship's weapon and sensor capabilities against enemy submarines of all types, patrol craft armed with anti-ship missiles, and surface ships. The primary mission of the Seasprite is anti-submarine and anti-surface ship warfare, anti-ship surveillance and targeting, and anti-ship missile defence. Furthermore, it can serve in such vital capacities as search-and-rescue, medical evacuation, small boat interdiction, amphibious assault air support, personnel and cargo transfer, mine detection, gun fire spotting and battle damage assessment.

Flying in support of amphibious operations, the Sea Stallion helicopters of a ship like *Wasp* are also fully competent in such roles as search-and-rescue, aircraft recovery, artillery lift, and the support of forward refuelling and rearming points.

Still another variant of the current US Navy assault

THE SMALL CARRIERS

Left: A November 1943 view of the flight deck and bridge of the escort carrier, USS *Cowpens* (CVL-25).

Nearly a hundred CVE escort or "jeep" carriers were built in US shipyards during World War II, using the hull designs of merchant ships as their structural basis. While often derided by their crews as "combustible, vulnerable and expendable", the small carriers took on the tasks that allowed the larger fast carriers of the US and Royal Navies to get on with their major roles. Slower, and with a capacity for only about twenty-four aircraft, the jeep carriers provided a vital resource through their involvement in aircraft transportation, amphibious support, close air support, and anti-submarine warfare.

ship is represented by the USS *Inchon* (MCS-12), a modified *Iwo Jima*-class LPH redesignated as a Mine Countermeasures Support ship. Explosive Ordnance Disposal is the mission of *Inchon* which brings up to thirty helicopters to its assigned tasks.

There is the *Austin* class of LPD, eleven of which are in the active USN inventory. They are *Austin* (LPD-4), *Ogden* (LPD-5), *Duluth* (LPD-6), *Cleveland* (LPD-7), *Dubuque* (LPD-8), *Denver* (LPD-9), *Juneau* (LPD-10), *Shreveport* (LPD-12), *Nashville* (LPD-13), *Trenton* (LPD-14), and *Ponce* (LPD-15). These sturdy little ships can range 7700 miles at 20 knots and are notable for their unusual tall twin funnels amidships. The *Austins* each host up to six Boeing CH-46D/E Sea Knight helicopters.

In addition to the above classes of assault ships, the USN operates two ships of the *Blue Ridge* class (LCC), each supporting a Sikorsky SH-3G Sea King helicopter; eight ships of the *Whidbey Island* class (LSD), which can each support two Sikorsky CH-53D Sea Stallion helicopters — as can its four ships of the *Harpers Ferry* class (LSD-CV).

Commissioned in September 1998, HMS *Ocean* (LPH01) is the first of the Royal Navy's new breed of assault ships capable of rapidly landing an assault force by helicopter and landing craft. With a hull design based on that of the current *Invincible*-class RN aircraft carriers, *Ocean* carries a crew of 255, with an air crew of 206, and 480 Marines. She is able to accommodate an additional 320 Marines on an emergency basis. She can transport and sustain an embarked military force of up to 800 men with their artillery, vehicles and stores. She can support twelve EH101 Merlin and six Lynx helicopters, and can carry, but not support, twenty Sea Harriers.

Equipped with the BAE Systems ADAWS 2000 combat data system, the command system of *Ocean* is compatible with the ships of the Royal Navy's front line fleet. Her weapons include four Oerlikon/BAE twin 30mm guns and three Raytheon/General Dynamics Phalanx Mk 15 close–in weapon systems. She is fitted with two decoy counter-measures

systems and is equipped with the BAE Systems Type 996 air and surface search radar, as well as two Kelvin Hughes search and aircraft control radar systems. The *Ocean* has a top speed of 18 knots and a range of 8000 miles at 15 knots. Early in the twenty-first century she will be joined in service by two other new assault vessels, HMS *Albion* and HMS *Bulwark*.

Some called them "Woolworth Carriers" . . . the American-built World War II Lend–Lease escort aircraft carriers of the *Attacker* class, including *Battler, Hunter, Chaser* and *Pursuer,* all of them products of the Ingalls Shipbuilding Corporation of Pascagoula, Mississippi. Lovely of line they were not, but neither were they cheap imitations of the real thing, though their Royal Navy crews found that certain adjustments were necessary in the way things were done aboard these new CVEs.

The routine and methods of the US Navy were incorporated into the design philosophy and construction of these ships in many areas, including accommodation for the ship's company, food preparation and service, and the operational spaces. In the working areas, these ships were relatively roomy compared to the layouts of most traditional British warships of the time. The hangar deck, for example, was spacious and of a kind of "open plan" concept, against that of the British-designed fleet carrier which utilized double hangars that were not as open and airy, or as user-friendly a workspace. For the recreation of the crews, the hangar area housed a full-size projection screen which could be lowered to turn the space into a cinema, and in hot weather the space could be made cooler and more comfortable when the after lift was lowered.

In British-built warships then, ratings slept in slung hammocks. In the CVEs, three-level bunks arranged in units of six, separated by narrow gangways, were positioned in deck spaces well below the hangar deck. When not in use, the bunk units could be folded up and secured to provide additional space. Overall though, the CVE crews, for the most part, considered conditions in their spaces to be claustrophobic and unhealthy.

In a tradition that is reflected in the design of today's supercarriers, bunking areas for ratings included a small recreational space with a few chairs and tables which were used for card games, reading, letter writing and relaxing. Petty Officers, the next level up in the pecking order, did not fair much

better than the ratings in accommodation. Officers, however, were afforded two-berth cabins similar to those in carriers today like the Royal Navy's *Invincible*-class, and the US Navy's *Nimitz*-class ships. They were equipped with a spacious wardrobe, a personal safe and a desk, as well as a two-bunk sleeping unit and additional storage facilities.

Operating departments functioned with such facilities as a parachute packing and drying room,

Below: A fine study of an American escort carrier during World War II. This one with F4U Corsair fighters on her forward flight deck.

complete wood and metal workshops, and a well-equipped sick bay. Unlike the Royal Navy system for food preparation — where duty cooks gathered food stuffs from the galley, made them ready on their wooden tables and then brought them back to the galley for the actual cooking — the CVEs were set up for most dishes to be steam-cooked in the galley. The meals were served and consumed cafeteria-style in a central dining room. Ratings ate their meals from sectioned metal trays, just as American sailors did. As on today's carriers, Petty Officers had their own mess, and officers dined in their own wardroom. There was also a NAAFI canteen and soda fountain for the men. Other facilities on the CVEs included a laundry and a barber shop.

The escort carriers of the *Attacker* class were essentially simple, utilitarian vessels of largely prefabricated and welded construction. They were powered by single shaft geared turbines and were capable of 17 knots maximum speed. At 492 feet in length, with a 69 foot beam, they carried four double 40mm and fifteen single 20mm anti-aircraft guns, and two 4-inch anti-aircraft guns. By 1942 the men and women of the Ingalls yards were assembling one such CVE a month. The journey from the shipbuilder's yards on the Gulf of Mexico, to deployment with the fleet of the Royal Navy, was a long and complicated one. The time it took was a cause of concern among some US Navy officials who were anxious about the delay in getting the CVEs into action with the Royal Navy against the U-boats of Grand Admiral Karl Dönitz. Delays of six months were common, with up to eight months for escort carriers constructed on the west coast of the United States. Such vessels had to go through initial sea trials before they could be released for the long trip through the Panama Canal to the east coast port of Norfolk, Virginia, where they would be loaded to absolute maximum capacity with new fighter and other aircraft destined for US units in Britain and North Africa. Frequently they would

wait weeks for the assembly of a convoy in which to make their first crossing of the Atlantic. Often they would stop at Casablanca to off-load aircraft before proceeding to the UK, adding as much as three more weeks to the sailing.

When the CVEs finally reached Britain, they would then spend up to eight additional weeks in dockyards for modifications to ready them for the rigors of combat. The British required a number of significant alterations, including the lengthening of the flight deck to accommodate some of their aircraft like the Swordfish, which was not able to launch using the American catapult system and needed additional take-off distance. Further British modifications were made to many of the Lend–Lease escort carriers to give them capabilities beyond the anti-submarine role the Americans had understood to be the intended use of the ships by the British Admiralty. The various delays were brought to the attention of Admiral Ernest King, the US Navy Chief of Staff, by members of the Allied Anti-Submarine Survey Board in August 1943. They found the lengthy delays unacceptable and recommended that King retain five out of the next batch of seven Lend–Lease escort carriers for use by the US Navy. The British Admiralty explained to King about the congestion caused by the workload in UK dockyards and how it was getting worse with new CVEs that were being delivered from the US west coast, as well as from the Gulf Coast, and about Royal Navy manning problems for the new ships. Persuaded by the Admiralty explanation, and aware that the US Navy was probably even less able at that point to man the CVEs in question, Admiral King came down on the side of urging the British to find ways to reduce their delays, which they pursued diligently.

Many fascinating and compelling names were assigned to the CVEs of the Royal Navy. Names that conjured images of roles they would play and actions in which they would engage. Names like *Searcher, Stalker, Chaser, Hunter, Pursuer, Ravager,*

Accidents would have been more common if boiler room crews had not done unbelievable tasks to provide full power for the needed "wind over the deck". Re-bricking the boilers was essential for full power. Putting the ship out of service for routine overhaul was impossible. The boiler room crews did this hot, hazardous task at sea, even during the Battle of Midway.
— Jack "Dusty" Kleiss, former US Navy pilot

Left: The escort carrier USS *Charger* (CVE-30) in January 1944.

Below: A TBF-1 Avenger torpedo bomber takes off from an escort carrier during World War II.

Attacker, Battler, Fencer, Striker, Patroller, Ruler, Reaper, Puncher, Arbiter and *Smiter.*

The US Navy escort carrier *Gambier Bay* (CVE-73) was the only American aircraft carrier to be sunk by gunfire in World War II. It happened during the decisive Battle of Leyte Gulf, 22–27 October 1944, when 321 ships and 1996 aircraft — more ships and planes than in any other battle in naval history — engaged in an action that put an end to the

Japanese fleet as an offensive force in the war.

The war was going badly for the Japanese in the summer of 1944. It was clear to them that the Americans were preparing to invade the Philippines and it was imperative that the Japanese retain possession of the island group. To lose it would mean the loss of their oil supplies and other vital raw materials from Sumatra, without which they could not survive and continue their war effort. The Japanese military believed that the US invasion of the Philippines would come through Leyte Gulf, and it was there that they would fight what they believed would be the greatest of all naval battles.

The *Gambier Bay* was built by Kaiser Shipbuilding at Vancouver, Washington. Her keel was laid on 10 July 1943. She was named after a small bay on the coast of Admiralty Island, Alaska, and was commissioned on 28 December at Astoria, Oregon. After commissioning, she went to San Diego on her shakedown cruise and on 7 February 1944 sailed for Pearl Harbor and the Marshall Islands. In the Marshalls she flew off eighty-four replacement aircraft for the US Navy fleet carrier *Enterprise* (CV-6), before returning to Pearl where she picked up damaged planes and brought them back to the States for repairs. In this, her initial Pacific cruise, she had developed a major vibration problem and was put in drydock at San Diego to correct it. By late March the *Gambier Bay* was back at sea in support of US Marine aviators in their carrier qualifications with F4U Corsairs. Then, on 1 May, she left San Diego for the Marshalls to be part of Task Group 52.11, carrying her own squadron, VC-10. The task group was preparing for the coming invasion of the Marianas Islands, whose capture was the key to obtaining essential airfields for basing the US B-29s that were to bomb the Japanese home islands in the final year of the war.

During the Marianas invasion, the carrier's squadron was kept busy flying close air support for the US Marines in their landings on Saipan and Tinian, accounting on 17 June for many

Japanese aircraft. The next stop for the ship was Guam where her pilots again flew air support sorties. More amphibious attack support followed at Peleliu and Angaur in the Palaus in September, and then, in company with *Kitkun Bay* (CVE-71) and four destroyers, the *Gambier Bay* escorted transports and amphibious landing ships to Leyte Gulf, as part of the force that would be involved in the coming invasion of the Philippines. Both *Kitkun Bay* and *Gambier Bay* were to be part of Admiral Thomas L. Sprague's Task Group 77.4.3, code-named Taffy 3. Other US Navy CVEs in this task group were *Kalinin Bay* (CVE-68), *Fanshaw Bay* (CVE-70), *St. Lo* (CVE-63), and *White Plains* (CVE-66).

It was during the morning of 25 October that Taffy 3 encountered a massive force of Japanese cruisers and destroyers off Samar Island. Captain Walter Vieweg, skipper of *Gambier Bay*, was awakened at 2.30 a.m. by his communications watch officer who informed him that the battle of Surigao Straights was under way. He ordered all of his uncommitted Avenger aircraft loaded with torpedoes to be ready to participate in the action. By 7 a.m. one of the task group's anti-submarine patrol planes reported that a Japanese force of four battleships, eight cruisers and thirteen destroyers, was only 25 miles north-west of the US task group. About this time, salvoes of large-calibre shells began to fall in the centre of the American ship formation, which was heading in a southerly direction. The US ships quickly turned east into the wind, in order to launch their aircraft. Captain Vieweg ordered all planes on the deck launched immediately in his concern that they would be lost should an enemy shell cause a fire on the flight deck. All ten fighters and seven torpedo bombers took off safely.

With the departure of the planes that had been on the flight deck, Vieweg began bringing planes up from the hangar deck. The task group, though, was changing course to a more southerly heading, and the relative wind over the deck of *Gambier Bay* was now minimal, theoretically insufficient for the launch of fully-loaded and gassed torpedo planes. Some of the aircraft, therefore, had to be stripped of their crews and jettisoned for the safety of the ship, as they could not be safely launched in the inadequate wind. The enemy salvoes were now falling much closer to the carrier.

By 7.30 a.m. the situation had become chaotic, with all of the US ships making smoke for their protection, the destroyers busily attacking the Japanese fleet, US planes (many of them without bombs or with inappropriate loads for the present task), making repeated air attacks on the enemy ships, continuing salvoes of heavy shells falling on and among the American vessels, and a lashing rain squall in full fury.

The rain stopped at 8.30 a.m. and, by which time all of the US Navy destroyers and destroyer escorts that had left the carrier formation to attack the enemy ships, were fully engaged in battle and smoke-screen laying. At this time both *Kalinin Bay* and *Gambier Bay* were situated on the windward side of the carrier formation, exposed to the gun fire of the nearest three or four enemy cruisers. The American smoke did provide some minimal protection for the two carriers against the fire of the more distant Japanese battleships.

The enemy cruisers continued to rain extremely heavy gun fire onto *Gambier Bay*, although much of it to this point had been relatively inaccurate and the salvoes rather widely spaced in time, with up to 1 1/2 minutes between them. The lag-time between salvoes enabled *Gambier Bay* to dodge many of them, a tactic that Captain Vieweg was able to continue for more than half an hour. But the distance between *Gambier Bay* and her adversaries was steadily decreasing and, when the Japanese cruisers had closed to within 10,000 yards of the escort carrier, the accuracy and intensity of their gun fire improved dramatically. *Gambier Bay* then received shells which exploded

Captain "Bowser" Vieweg was a fine gentleman, quiet, not at all flamboyant. During the final attack on the ship, while men were fighting fires, dragging hoses, and manning the guns, a hit disabled our catapult, and another jammed our elevator, bringing flight operations to a halt. This completed my job as I had no other general quarters battle assignment. I ran forward to assist any damage control party that I could, and saw enemy ships bearing down on us from dead ahead. We were being fired on from the port side. I looked up at the bridge and saw Captain Vieweg calmly giving orders to the helm. I later learned that the ship was being steered towards the shell splashes. Whenever a shell burst ahead, he would steer for the burst to throw the enemy gunners off, all the while looking at the enemy through his binoculars.
– Morris Montgomery, formerly with VC-10, USS *Gambier Bay*

on her flight deck and near her engine room, the latter causing major flooding and the immediate loss of one engine, reducing the ship's speed to just 11 knots. The little carrier began to fall back from the rest of the formation.

Now the enemy salvoes were scoring hits with consistency and starting to cause significant damage to *Gambier Bay*. The volume of firing by the Japanese increased and, for the entire hour from the first enemy shell hitting the carrier at 8.10 a.m., until 9.10 a.m., the shells were striking effectively at the rate of at least 50 per cent. Many shells hit just below the water line causing more flooding and the loss of the after engine room. The ship was now essentially dead in the water, with no water pressure to fight the small fires being continually set off by the striking shells. The small fires quickly became large fires and by 8.50 a.m. it was clear to Captain Vieweg that his ship could not be saved. He ordered it abandoned as the intensity of the enemy gun fire increased still more. Even as *Gambier Bay* rolled over at 9.04 a.m., and until she sank about six minutes later, the Japanese cruisers continued to pour heavy shells into her. In addition to *Gambier Bay*, her sister ship *St Lo*, the destroyers *Hoel* and *Johnston*, and the destroyer escort *Samuel B. Roberts,* were sunk on the morning of 25 October. The CVEs *Kalinin Bay*, *Kitkun Bay*, *Fanshaw Bay* and *White Plains* were badly damaged. The carrier *Princeton* was also lost after sustaining two bomb hits from a Japanese Judy aircraft, causing a fire and massive explosion which resulted in the ship having to be scuttled.

The Battle of Leyte Gulf cost the Imperial Japanese Navy four aircraft carriers, three battleships, six heavy cruisers, four light cruisers, eleven destroyers, one destroyer transport and four submarines. The US Navy lost one light aircraft carrier, two escort carriers, and three destroyers. Aircraft losses numbered 150 for the Japanese against 100 for the Americans. The Japanese lost approximately 10,000 men; the Americans lost 1500. The US Navy's

234

A SELECTION OF NAVAL AVIATOR CALL SIGNS

ASTRO
BOA
BOZO
BUG
BULLDOG
BUNS
BWANA
CHEF
CHIEF
CIRCUS
COBRA
CURLY
DELBOY
DISNEY
DOC
DOGGO
DOMINATOR
DUDE
EGG
EIGHT-BALL
FESTUS
FINGERS
FLASH
FLEX
FLOUNDER
FLY
FOOD
FUDGE
GABBY
G-MAN
HACK
HOLLYWOOD
HORSE
HOSE MONSTER
HOSER
ICEMAN

greatest ace of the war, David McCampbell, who commanded Air Group 15 from the carrier *Essex*, had already accrued twenty-one confirmed victories by the morning of 24 October when he and six other F6F Hellcat pilots launched to intercept a large force of Japanese dive bombers and torpedo planes being escorted by forty Zero fighters. McCampbell and his wingman, Roy Rushing, took on the enemy fighters and in the encounter McCampbell downed nine, while his wingman accounted for an additional five, becoming an ace in a day. For his achievement, David McCampbell was awarded the Congressional Medal of Honor. By the end of the war he had been credited with thirty-four enemy aircraft destroyed in the air and twenty on the ground.

Before leaving his ship, Captain Vieweg made his way down from the bridge through the island structure in an attempt to be sure that all of his men had abandoned the carrier. As he descended in the island interior, which by this time had become a "funnel" of smoke and hot gases, he found his escape route barred. Just then, another enemy salvo slammed into the bridge, forcing him to continue down through the smoke to the flight deck where the inferno was such that he was unable to see. Groping along the side of the island, he found and fell into the catwalk, but soon managed to lift himself over the side. He fell 40 feet to the water, recovered his senses and realized that his ship was about to roll over on him. He quickly put 100 yards between himself and the vessel. Captain Vieweg and about 150 men in a cluster of life rafts were picked up at 4.30 a.m. on 26 October. Approximately 700 men of the *Gambier Bay* survived and were rescued by 27 October.

Morris Montgomery, Lieutenant Commander, US Navy, Retired, is the grandfather of Lieutenant J.G. Shannon Callahan, whose comments appear in an earlier chapter, *Sea Change*. Montgomery served as the leading Chief Petty Officer of Composite

Squadron 10 which was embarked in the USS *Gambier Bay* during the battle off Samar, Philippine Islands, on 25 October 1944. "On the night before the ship was lost, one of the the two tractors on the flight deck somehow tumbled down the elevator opening and landed on a fighter on the hangar deck. The damage was such that the plane required an engine change and the maintenance crew worked all night changing the engine.

"Our chief ordnanceman Andy Andrews, who bunked near me, was awakened early in the morning, and I knew something was up when I heard about loading torpedoes. Our planes had previously been loaded with general purpose bombs, for a land bombing mission the following day. To reload them with torpedoes meant bringing them below to the hangar deck and de-arming and de-fuelling them, a time-consuming operation. At about 6.30 a.m. Chief Horten, the aircraft maintenance chief, and I went to the flight deck to relieve the maintenance crew who were preparing to ground-run the new engine that had been installed in the fighter that had been damaged on the elevator. They went to breakfast, Chief Horten got into the cockpit, and I stood by with a CO_2 fire bottle. The engine would not run on its engine-driven pump, only on the electric booster pump, which was used mainly for starting and then switched off. We decided to complete the ground-run using the booster pump and change the engine pump later during the post-engine-run inspection.

"At that moment the ship was surrounded by rain squalls, but was not actually in one. During the engine-run, I noticed one of our morning anti-submarine patrol planes approaching the ship and 'signalling' with its wings. I then saw a big splash in the water about 100 yards to port. At first I thought that the aircraft was in trouble and had jettisoned its depth charges. Then I saw a Japanese battleship emerging from a rain squall on our port horizon and heard the roar of its guns. The sound of the enemy shells passing overhead reminded me of a train passing a crossing at great speed.

"Things began to happen. People were running around, 'general quarters' was sounded and the ship began to make screening smoke and take evasive action. Pilots began scrambling for any available aircraft. One made for the fighter with the new engine, and we told him that the booster pump was all he would have. He acknowledged and was subsequently launched. New aircraft engines had to be operated at minimum power for at least ten hours, with no combat or military power settings. The pilot eventually experienced engine failure and had to ditch.

"I could see a formation of enemy ships to our port side, seemingly at point blank range. It was a mad scramble to get our planes launched. Some were not completely re-fuelled after the switch of ordnance, and I know that some had *no* ordnance. At least one was catapulted without the crew, just to get it off the ship. Either it had been 'down' mechanically or had not been re-fuelled. The fuelling system was then secured, as was normally done at 'general quarters'. 'Dead' planes and aviation gas in the fuel risers are definitely a disadvantage in a situation such as the one we were experiencing.

"Each time the ship received a hit, footing became difficult to maintain and the generators were knocked off line, cutting the power and lights momentarily. One hit jammed our elevator, trapping those planes that were on the hangar deck.

"Once flight operations had ceased, my job on the flight deck was finished and I had no other assignments. I went to the ready room. There were several pilots there and I asked 'What the hell was happening?' One pilot said that the fleet must have all been sunk and that we were all that was left. He thought that we had not been truthfully informed as to how the war had really been going for us.

"I grabbed a Mae West life jacket and left to help out wherever I could. Men were fighting fires everywhere. I remembered that I had not put the flame cover over my bunk that morning, so I went below to the CPO berthing space. The quarters contained an emergency first aid station which also served as our writing table, and it doubled as an operating table, having special lights mounted above it. When I entered, the quarters were ankle-deep in water, possibly from a broken main. A medical crew was working on an injured man, and the lights were periodically blinking off and on. They had battle lanterns in use and it was an eerie sight.

"I kept a picture of my wife in a pocket-sized copy of the New Testament under my pillow. I put the Bible in my shirt pocket, spread the flame cover over my bunk and went back to the flight deck. I attempted to assist the damage control crew in 'Officer Country' below the flight deck up near the bow. They had all the help they needed, so three other men and I headed further up the bow when the ship received another hit. The men ahead of me seemed to disintegrate, and I was knocked down. I got up and headed back towards the Officers' quarters where I met Lieutenant Bell. He was getting ready to abandon ship, and he ducked back into his room and offered me a drink from a bottle he had. We all started towards the port railing on the bow.

"The water was filled with men. I never did hear the word to abandon ship, but I knew that the end was near because the ship was dead in the water and beginning to list. I overheard the leader of one of the damage repair parties say that the magazines were being flooded. I went to the railing and then loosened the straps of my battle helmet. Training lectures had taught me to do this to avoid being injured upon entering the water. I now had a Mae West, a life belt, and a kapok life jacket which I had found on the bow. Shells were hitting, leaving various colors from their dye markers on top of the water. I went over the side.

"I hit the water and surfaced near one of our pilots. He had a mattress so I joined him. I never

JAMBO
JAWS
KNUCKLES
LIGHTNING
LOCK & LOAD
LONER
MACK
MISSING
MOE
MUSTANG
NO-SHOW
ORGAN
OTTO-DOG
PIGLET
PILLSBURY
PLAYER
RAT
ROADRALPH
ROCKY
SECKS
SLIDER
SLIM
SMITTY
SMOKE
SLODGE
SPOOK
SPY
SQUIRE
STRIKER
SUNSHINE
SWEEPER
TERMINATOR
TEX
THUMPER
THUNDER
TURTLE
TYPHOON
VIPER
WHEEZER
WHIZZER

saw Lieutenant Bell again. I was glad I had gone over the port side as the tide was pushing against the starboard side. We were being carried away from the ship unlike those men who had gone over the starboard side and were forced to follow along the ship's hull and drift off the stern. As we drifted away we were constantly splashed by water from nearby shell hits. A few yards off the stern were several men in one of the large donut-type floats which had water and rations stored in them. They were waving and yelling for people to come over to them as they had plenty of room. We began paddling in their direction. About then there was a terrific explosion, probably on the hangar deck, because the elevator (which had been jammed) came flying out of the ship and landed near us. I am certain that some men must have been struck by it when it hit the water. The float that we were heading for must have received a direct hit for as we neared it there was a big splash, and then it wasn't there anymore. We continued to drift away from the ship and shells continued to fall near us. We began to think that the enemy gunners were using our mattress as a marker. Of course they were not, but one's reasoning in such situations is not always the best. So we abandoned the mattress and swam away from it. We soon came upon some men who were trying to spread out one of the ship's large floating nets. Such nets are unwieldy, requiring men on all sides of it, but it makes a very effective piece of life-saving equipment. By keeping the net spread out, those in the middle of it could remain about waist deep in water and have their feet supported. We put the more seriously wounded in the net. We spotted a man floating near the net and swam over to him. He was dead. He could have been a mess cook or a striker, because he was young and was wearing whites and was non-rated. We took his I.D. tags and gave them to the senior officer in our group. The *Gambier Bay* rolled over bottoms-up revealing many holes in the bottom before it sank.

"We could see a Japanese cruiser which appeared to be dead in the water. Smoke was billowing from its after section, and a Japanese destroyer was trying to assist it. At about 1300 hours we saw many planes flying in the direction of the Japanese fleet. We identified them as ours. We all shouted and waved our arms, and several of us yelled 'they went thataway'. We were in pretty good spirits at that time. Later we thought how fortunate we were that the Japanese force had gone before our planes arrived. Otherwise, we might have been hit during the bombing attack. I remember seeing the Japanese cruiser burning just before dark. It may have sunk during the night as there were several underwater explosions.

"It was very cold at night, especially if any part of your body became exposed to the air. The water was a bit rough and we were constantly being drenched with salt water. We tied ourselves together with the attachment strings on our life jackets. That way, if you fell asleep or dozed off, you wouldn't drift away. One of the wounded men had a large piece of metal stuck in his back, apparently wedged so tightly that he was only bleeding slightly. The doctor said not to pull it out or else he might bleed to death. We took turns holding his head up out of the water and trying to comfort him. On the second day, as I was holding him, he began to talk about things back in his childhood. As the day wore on he became incoherent. Finally, he went into a coma and died, possibly from shock and internal bleeding. We took his I.D. tags and gave him a burial prayer. We then allowed his body to drift away and sink.

"We had about ninety men in and around our net. Everyone was getting thirsty and hungry. At one point I heard two men talking. I don't know if they were joking or temporarily out of their heads. One of them said that his father owned the bar around the corner and if the other one would go with him, he would set up the drinks. They both paddled around to the other side of the net and began to

Above right: Morris Montgomery in 2000. Below right: Bracketed by shell fire from Japanese battleships and cruisers, the escort carrier *Gambier Bay* (CVE–73) is mortally wounded off Samar in the Battle for Leyte Gulf on 25 October 1944.

drink salt water. The men near them promptly stopped them from doing so. Those of us who were not severely wounded and had control of themselves would watch for men attempting to drink salt water and keep others from swimming away. During that time one's mind would play tricks and you would see objects like mirages, or imagine things. I saw a 'fleet of rescue ships', or so I thought, until someone shook me. In the night as I was dozing, one of the men near me began pounding me on the head and screaming that I was trying to get ahead of him in the chow line. We prayed frequently for rescue, food and water.

"Later, a few of us spotted something bobbing on

our horizon. It turned out to be Chief McArdle towing two wounded men. One of them was Forest Khort and the other was Denard, both from our squadron. They had a small water breaker and part of a can of emergency rations . . . malt tablets. We helped them to our net and rigged a drinking hose from one of our Mae West inflation hoses to a spigot on the water breaker. We all received a small sip of water and one malt tablet. A 2nd Class Storekeeper from our group took charge of the water and rations and did the rationing. We saw to the wounded men in the middle of the net first. I don't claim to be any more religious than the next man, but it was at this point that my faith in God

was most profound, and I then had faith that we would be rescued.

"We lost several men, mostly from just drifting away or taking off on their own. At first we would swim out and bring them back, but near the end of our time in the water, we didn't have the strength to do this any more. We learned later that some of the men who had left our group had eventually floated into another group.

"Most of us were badly sunburned from the glare off the water. On the second day we found a floating drum of oil and we smeared some of the oil on all of us, but it was a little too late. My lips were raw and bleeding and felt like they stood out a foot in front of me. At that point my Mae West had lost most of its air and the inflation hose had clogged with congealed fuel oil. My kapok life jacket was partially water-logged. On the first day in the water I had given my life belt to someone who needed one. One of the ship's warrant officers had found a wooden gangway post and, realizing the condition of my life-saving equipment, I joined him, each of us clinging to an end of the post. It was good to be on one end of that piece of lumber with him. We had good manoeuvrability and were supported somewhat. To be able to support one's dangling weight just a little bit was a tremendous relief. Our life jackets were becoming water-logged, causing us to sink further into the water. As the seas got heavier, the float net kept trying to roll up, and it was a real chore keeping it straightened and spread out. The sea state was becoming worse and ominous clouds were forming. It began to worry us. We stayed near the net, our legs quite numb from dangling in the water.

"On our second night in the water we spotted what we thought was Samar Island in the distance and had to decide whether to risk the possibility of drifting ashore there. We reasoned that it was enemy-held, but we decided that it was our best chance for survival and we thought that some of the wounded men would not survive another day

in the water, so we began to try to control our drift towards the island.

"Just before daybreak we saw some short sweeps of searchlights, on and off. Someone thought that they might be Japanese torpedo boats. We finally agreed to try to indicate our position. We yelled and blew the whistles attached to our life jackets. We were suddenly engulfed in the beam of a searchlight and I could see an American sailor near the light source. It was the most welcome sight of my life. The men on board the craft began tossing out heaving lines and putting out litter-type stretchers wrapped with life jackets for bouyancy for the wounded. I was hauled up on the deck and my oil-soaked clothes were cut away. I said, 'I'm OK. I can make it', and when they released me my legs just folded up and down I went. They carried me below. Naked, oily men were stacked up like a fresh-caught load of fish. Apparently, they had picked up other groups of survivors before us. Someone gave me a small sip of water and, a minute of so later, another one, and so on. My lips were so swollen and painful, and my tongue so raw and ulcerated from constantly spitting out the salt water, that I actually cried when the cup even came close, but my desire for water was stronger than the ensuing pain. Other than a little soup, the first food we were given was pancakes. Even though I was not fond of Navy pancakes, these looked and smelled delicious, and I very much wanted to eat them. I became quite frustrated, but the pain was just too great and I had to forego that meal.

"Eventually we made it to Leyte Harbor, which was filled with troop transport and utility ships. I was transferred to an LST (Landing Ship Tank) that had been converted to an emergency hospital ship, and was examined, treated and released as an ambulatory patient. I was suffering from severe shortness of breath, diagnosed as resulting from concussion. I had various cuts and bruises, severe sunburn and other effects from exposure. I was one of the lucky ones."

Right: The *Belleau Wood* (CVL-24) afire on 30 October 1944 after an attack by a Japanese kamikaze aircraft. The fire killed fifty-six men and destroyed twelve planes. *Belleau Wood* was forced to withdraw to the Ulithi anchorage for repairs.

AIR STRIKE
PERSIAN GULF

The thund'ring cannon now begins the fight, / And though it be at noon, creates a night. The air was soon after the fight begun, / Far more enflam'd by it, then by the sun. / Never so burning was that climate known, / War turn'd the temperate, to the torrid zone.
– from *The Victory Obtained by Blake over the Spaniards, in the Bay of Santa Cruz in the Island of Teneriffe, 1657* by Andrew Marvell

Right: VF-84 squadron maintenance personnel resting on the wings of an F-14A Tomcat tied down on the flight deck of the nuclear-powered carrier USS *Theodore Roosevelt* (CVN-71). The *Roosevelt* was on station in the eastern Mediterranean in support of Operation Provide Comfort, a multinational effort to aid Kurdish refugees in southern Turkey and northern Iraq in May 1991.

IT WAS SET TO BE the first daylight strike of the Gulf War, the Allied Coalition effort to force the invading Iraqis back from their incursion into neighbouring Kuwait. F/A-18C Hornets from the US Navy carriers *Saratoga* (CV-60) and *John F. Kennedy* (CV-67) were to launch the daylight phase of an ongoing strike series to be conducted round the clock. The night before, 16 January 1991, the black sky over Baghdad was rent with streaks and dotted lines of neon-like anti-aircraft artillery fire. The triple A was in response to a massive first-strike by several hundred Coalition aircraft, on airfields and other facilities around the city. Iraqi surface-to-air missiles flew at the attackers, and a few interceptors of Saddam Hussein's air force rose to the occasion. In the evening's action three Iraqi MiG-29s and three Mirage F1s fell to the weaponry of US Air Force F-15C Eagles. The flak and SAMs accounted for some Allied aircraft losses; the Iraqi interceptors remained but a threat.

The next day the Navy continued the activity. Lieutenant Commander Mark Fox was flying from the *Saratoga* in one of the spare aircraft, airborne in case any of the other Hornets on the raid should develop a problem and have to abort. In fact, three squadron aircraft had to return to the carrier, unable to complete their assignment, and Fox moved into the line-abreast formation to join the mission.

The four F/A-18s proceeded on course towards their airfield target, H-3, in western Iraq. They were heavy, each of them loaded with four Mk 84 2000-pound bombs. The Hornets were being monitored from above by an AWACS E-3B Sentry aircraft from the King Khalid airbase near Riyadh, and an E-2C Hawkeye of VAW-125 from the *Saratoga*.

Mark Fox had prepared well for the attack he was running in on. He had set and checked the required switches for the drop. As the *Saratoga* Hornets approached the target from the south, Fox began hearing radio calls from aircraft of the *Kennedy* strike force Hornet pilots who had sighted a few Iraqi MiG-29 Fulcrums following them as they left

242

the target area after releasing their ordnance. In seconds two USAF F-15C Eagles appeared from a higher flight level and fell upon the MiGs, disposing of them and their potential threat to the heavily-loaded Hornets.

Now the Hornets were less than 35 miles from their bomb release point, when the Hawkeye called them with a "bandit alert". More MiGs were in the area and were believed to be heading for the F/A-18 flight as it continued towards H-3. The situation became serious when the Hornets received a second call from the Hawkeye, telling them that the MiGs were 15 miles dead ahead. At that moment, both Lieutenant Commander Fox, and Lieutenant Nick Mongillo in a nearby Hornet, elected to reset their weapons systems to "missiles" for possible air-to-air engagement. When the four Navy pilots drew within ten miles of the bandits, three of them got "missile-lock" on the two enemy aircraft which were MiG-21 Fishbed fighters. Fox had the MiG on the right; Mongillo took the other one. Radar missile-locks were then also achieved on the MiGs by the other two Hornet pilots.

The MiG fighters approached the F/A-18s head-on at supersonic speed and did not waiver. Fox was the first to fire, sending a Sidewinder towards his MiG. For good measure he also fired a Sparrow missile at his target. The MiG became a fireball, absorbing both missiles, but it was the Sidewinder that had destroyed it. Within seconds of Fox's kill, Mongillo loosed a Sparrow at the MiG in his lock, splashing it as well. Though heavily loaded with bombs and not ideally set up for aerial combat, the two F/A-18s had made their kills and now joined up with the other two for the bomb run. They attacked the airfield target and turned south again on a heading for the Gulf and their carrier. Two towering columns of black smoke marked the locations on the desert floor where their MiGs had impacted.

The incident may have been the first when fighter/attack airplanes carrying 8000 pounds of

The only time you have too much fuel is when you are on fire.

Learn from the mistakes of others. You won't live long enough to make them all yourself.

You start off with a big bag of luck and an empty bag of experience. The trick is to fill the bag of experience before you empty the bag of luck.

In the Persian Gulf a US Marine mans a .50 calibre machine gun aboard a Marine Medium Squadron 161 helicopter.

245

Above: In a USN squadron ready room briefing. Right: A Zippo lighter with the insignia of VF-211, which was embarked in the *Stennis* in November 1999. Far Right: HMS *Illustrious* (left) alongside the 102,000-ton USS *John C. Stennis* (CVN-74) on station off Bahrain in winter 1999.

246

Hydraulic catapults designed by the Naval Aircraft Factory in Philadelphia were used in US Navy aircraft carriers until the 1960s. They did not give smooth, uniform pushes to the aircraft, and they lacked the power needed for new models of aircraft that were coming to the fleet. The same Naval Aircraft Factory designers built a "controlled explosion" prototype catapult (which used oil, air and water) to be installed in the new USS *Enterprise* (CVN-65). The catapult was behind schedule and had lots of problems. My catapult design engineers in the Bureau of Aeronautics didn't like it. We sent for Colvin Mitchell, the designer of the steam catapult. Colvin's design was great, but there was one problem. When you ask for more steam on a nuclear ship, you can get LOTS more steam quickly, but the pressure drops. We decided that the Navy knew steam and could surmount any problem with it. Mitchell's design was chosen. Steam catapults work beautifully.
– Jack "Dusty" Kleiss, former US Navy pilot and Bureau of Aeronautics representative

bombs engaged and shot down other fighters, and then were able to continue on and complete the bombing attack. Lieutenant Commander Fox knew, however, that if the MiGs had been able to get behind his flight, the Hornets would have been compelled to jettison their bombs in order to take on their adversaries. In the only confirmed US Navy air-to-air victories over Iraqi fixed-wing aircraft in the Gulf War, the Hornet proved the really special plane most Navy pilots had long believed it to be.

"Flight operations consisted of various missions for each specific tactical aircraft. Fighter missions varied from the boring grid search looking for submarines, to actual combat. Obviously, missions are dependent on the world political picture — news we rarely knew.

"Standing 'alerts' is part of carrier life. We fighter pilots had three alert stages: 30, 15 and 5 [minutes]. The status refers to the amount of time required to be airborne and ready for combat. 'Alert 30' required us to stay within 1000 feet of the aircraft. In other words, don't leave the boat. 'Alert 15' meant being in full flight gear, sitting in the ready room. It also required getting an inertial navigation system alignment on the F-14 to allow for a 'quick align' should there be a need to launch. 'Alert 5' was 'stood' sitting in the aircraft, strapped in and ready to go. At times the aircraft was even hooked up to the catapult. The watches were stood in succession: 30, then 15, then 5, with each stage 2 hours long. At night, 'Alert 30' was 'stood' in your rack; 'Alert 15' in the ready room watching a video of anything even remotely entertaining, or sleeping in a chair. 'Alert 5' at night was 'stood' in the aircraft, either sleeping or reading a book. There was a friendly competition between the two sister squadrons to be the first off the deck, and in one dramatic incident a RIO, in order to be first off, took a bad INS alignment. The crew were

stationed on a combat air patrol a few hundred miles from the carrier, outside of radio range of the ship. In the course of their flight, they lost their alignment and bearings, and could not relocate the ship and were not close enough to contact the ship by radio. Eventually they ran out of fuel and were forced to eject. Luckily, they were picked up by another ship after spending a very long night in their life raft."
– Frank Furbish, former US Navy fighter pilot

When Kelly Kinsella was a high school student in the Washington DC area, she was interested in attending a college that had a good competitive swimming programme. In the middle of her senior year, she received a call from the swim coach at the United States Naval Academy, recruiting her to swim there. Her grandfather had been a Class of '35 Academy graduate, but that had been the extent of her involvement with the institution. Intrigued by the possibility of being accepted for enrolment in the Naval Academy, she applied and did very well there, so well that by the second semester of her senior year, she was entitled to select her choice of Navy career assignment areas; she chose aviation.

Following her 1993 graduation, Kelly reported to NAS Pensacola for API and Primary training. From there she selected the E-2/C-2 pipeline. Although at that point jets were open to women aviators in the Navy, she preferred the C-2 Greyhound or COD (carrier onboard delivery) community and has never regretted her decision. "I really enjoy my job and think that I get more out of flying people to and from the carrier than I would dropping bombs. My job may not be as glamorous as a fighter pilot's, or seem as important to an outsider, but I wouldn't trade the C-2 community for any other.

"The final test for the E-2/C-2 pipeline is to carrier qualify in the T-2 Buckeye. The first time one lands on a carrier is one of the greatest days in your life. It is an amazing experience to fly solo, in a four-plane formation with one instructor in the lead plane and two of your close friends rounding out the flight, out to a carrier. Circling overhead, you ask yourself if you are *really* supposed to land on that small ship down there.

"After winging, I went to the Fleet Replacement Squadron (FRS), which is responsible for training newly winged aviators in the aircraft that he or she will fly in the fleet. Again, qualifying at the carrier is the final check at the FRS, only this time you get to go at night as well. Night carrier aviation is what sets naval aviators apart from all other pilots. I am now in my second WESTPAC deployment, doing night landings. It takes a long time to become confident at night. To say that one ever becomes comfortable at night would be a bit misleading.

"I have been a part of VRC-30 since finishing the FRS in March 1997. During my time in this squadron, I will have done two WESTPAC deployments to the Arabian Gulf, and numerous evolutions off the coast of southern California. One of the nice parts of being a COD pilot is that we are always supporting one carrier or another. There have been very few months at the squadron when I did not go out to the ship to get some traps. I have flown all sorts of folks out to see the ship and that is what makes the job so much fun. I have met Cabinet members, foreign dignitaries, Hollywood stars and famous athletes. Famous or not, to see the look on anyone's face as they get ready to fly out to the ship, or when they have just had their first cat shot, reminds me that I have a rare job indeed. We sometimes get run a bit ragged, flying back and forth several times a day, but the enthusiasm of the people we take to the carrier is contagious and really makes the job enjoyable.

"I have obviously not known any different than the Navy of today, where women fly any and all aircraft. The COD has been flown by women longer than any other carrier aircraft because it is not considered a combatant aircraft. There have been many times when I have found myself to be the only woman in a room, but that can happen in almost any traditionally male-dominated job. I have been trained exactly like every other aviator, past and present, to do a job. The women who paved the way in carrier aviation did a remarkable thing and made it easier for my age group to follow. The men I work with who are my age have always worked with women, so it is not an issue. Some of the older generation are still getting used to having women in the squadrons, but so long as we are doing the job well, there's nothing they can say anymore about our presence. We are here and we are doing well.

"Being on a carrier with almost 5000 men, I would be lying if I said I didn't receive any different attention for being a woman. You stand out and that's just a fact. But again, the most important thing is getting the job done. I feel very fortunate that on this carrier [USS *John C. Stennis*] there are several female aviators. My last Airwing only had two female pilots — myself and my room-mate, an E-2 pilot. We felt we were under the spotlight quite a bit, but that comes with being the only two women. This Airwing has women in every type of aircraft. My room-mate is a HELO pilot, also on her second cruise. The two women next door to me were naval aviators before combatants opened up to women, flying A-4s in the Philippines and Puerto Rico. One is now an F/A-18 Hornet pilot and the other is an EA-6B Prowler pilot. They both have flown combat missions and are respected for their abilities as much as any pilot in their squadrons. Having two senior aviators like them on board also helps set the standard for the new women checking into squadrons. Their professionalism and talent set the tone for the Airwing. There are also two F-14 RIOs, two E-2 NFOs, two S-3 pilots and two other HELO pilots (in addition to my room-mate). So, we are well represented. The best part about being with all of these women is that they are typical

Snap back the canopy, / Pull out the oxygen tube, / Flick the harness pin / And slap out into the air / Clear of the machine.

Did you ever dream when you were young / Of floating through the air, hung Between the clouds and the gay / Be-blossomed land? Did you ever stand and say, 'To sit and think and be alone In the middle of the sky / Is my one most perfect wish'?

That was a fore-knowing; You knew that some day / To satiate an inward crave / You must play with the wave / Of a cloud. And shout aloud / In the clean air, / The untouched-by-worldly-things-and-mean-air, / With exhilarated living.

You knew that you must float From the sun above the clouds / To the gloom beneath, from a world / Of rarefied splendour to one / Of cheapened dirt, close-knit / In its effort to encompass man In death.

So you can stay in the clouds, boy, You can let your soul go onwards, You have no ties on earth, / You could never have accomplished / Anything. Your ideas and ideals / Were too high. So you can stay In the sky, boy, and have no fear.
– *Parachute Descent*
by David Bourne

Below: The nuclear supercarrier USS *Dwight D. Eisenhower* (CVN-69) in the Suez canal *en route* to her six-month Gulf deployment in 1992.
Below right: An F-14A of VF-1 over the 1991 desert oil well fires in Kuwait.

women who just happen to also be naval aviators. We are not here to prove anything to anyone or to be 'one of the guys'; we get along great with our fellow aviators and are doing the job the Navy has trained us to do. We are all good friends, and having the support of one another makes life at sea much more enjoyable.

"Following this cruise, I have chosen to take non-flying orders. The decision to not fly for the next three years was a difficult one, but at this point it is the right one for me. I am going back to the Naval Academy to work in the swimming department. I look forward to sharing everything I have learned with the future aviators, especially the women. It's a level playing field now and hopefully they are ready to take advantage of that.

I am ready to move on and have a family — my fiancé will still be flying for the Navy and I want to be home enough to start having children. I know I will eventually go back to flying, perhaps civilian flying or in the reserves. I don't think anyone can completely walk away from flying — it's too amazing. My dream is to fly seaplanes, either in the Northwest or in the Caribbean, depending on my family. I hope my kids will be proud to say that their mom was a carrier aviator who flew all over the world and met many fascinating people. How many people can tell their children they have done what I have been fortunate enough to experience? At times I look back and wonder how this all happened. I never set out to do anything so great, but I sure have had a ball."

Somebody said that carrier pilots were the best in the world, and they must be or there wouldn't be any of them left alive.
– Ernie Pyle

This is the transcript of a radio conversation between a United States Navy ship and Canadian authorities off the Newfoundland coast in October 1995. It was released by the US Chief of Naval Operations, 10 October 1995.

Canadians: Please divert your course 15 degrees to the South to avoid a collision.
Americans: Recommend you divert your course 15 degrees to the north.
Canadians: Negative. You will have to divert your course 15 degrees to the South to avoid a collision.
Americans: This is the Captain of a US Navy ship. I say again, divert YOUR course.
Canadians: No. I say again, you divert YOUR course.
Americans: This is the aircraft carrier USS *Lincoln*, the second largest ship in the United States Atlantic Fleet. We are accompanied by three destroyers, three cruisers and numerous support vessels. I demand that you change your course 15 degrees North. I say again, that's one five degrees North, or counter-measures will be undertaken to ensure the safety of this ship.
Canadians: We are a lighthouse. Your call.

Right: A crewmember of a Westland Sea King helicopter embarked in HMS *Illustrious* in 1999.

Picture Credits
Photographs by Philip Kaplan are credited: PK. Photographs from the author's collections are credited: AC. Images from the Tailhook Association are credited: TH. Images courtesy HMS *Illustrious* are credited: HMSI. Images from the US Navy are credited: USN. Images from the Imperial War Museum are credited: IWM. Images from the National Museum of Naval Aviation are credited: NMNA. Images from the National Archives are credited: NARA. Jacket front: TH/USN, Jacket back: TH/USN, Jacket back flap: Margaret Kaplan. Front endsheet: TH/USN, Back endsheet: TH/USN, PP2-3: TH/USN, P5: TH/USN, PP8-9: TH/USN, P11: HMSI, P13: AC, P14 all: AC, P15: TH/USN, P16: TH/USN, P18 all: TH/USN, P19: TH/USN, P20: courtesy Merle Olmsted, P21: TH/USN, PP22-23: HMSI, PP24-25: TH/USN, P26 top left: PK, P26 top centre: PK, P26 top right: USAF, P27 top: TH/USN, P27 bottom: Neil Mercer, PP28-29: courtesy Michael O'Leary, P31: TH/USN, P33: TH/USN, P34: AC, P35: NMNA, PP36-37: TH/USN, P38: Neil Mercer, P40: USN, P42: AC, P44: TH/USN, P47: HMSI, P48: TH/USN, P49: TH/USN, P50: courtesy Robert Elder, P51: Painting '*Hook Down, Turning Final*' © 1999 Craig Kodera/Mike Machat courtesy Craig Kodera/Mike Machat and Casey Law, P54: TH/USN, P55: Neil Mercer, PP56-57: TH/USN, PP58-59: Neil Mercer, P60 both: AC, P62: courtesy John Wellham, P63: HMSI, PP64-65: courtesy John Wellham, P66: USN, PP68-69: courtesy Henry Sakaida, PP70-71: USN, P73: courtesy Bill Barr, PP74-75: TH/USN, P75 margin: courtesy Michael O'Leary, P76: courtesy Alexander Vraciu, P78 margin: courtesy Michael O'Leary, P79: TH/USN, P80 both: NMNA, P82: courtesy Thomas Harris, P83 margin: courtesy Michael O'Leary, PP84-85: NARA, P86 all: TH/USN, P88: NARA, PP90-91: TH/USN, P93: courtesy Bill Barr, P94: TH/USN, P95: PK, P96: Bill Barr, P97: NMNA, P98 margin: courtesy Mrs George Gay, P98: NMNA, P99 top: NMNA, P99 bottom, both: TH/USN, P100: courtesy Wilbur Webb, P101 top: TH/USN, P101 bottom: Bill Barr, P102-103: PK, P105: PK, P106 both: TH/USN, P107: TH/USN, P109: TH/USN, P110: TH/USN, P111: Neil Mercer, PP114-115: TH/USN, P116: Bill Barr, P118: AC, P120: NARA, P123: AC, P124-125: TH/USN, P126 both: NMNA, P128 top: TH/USN, P128 bottom left: Bill Barr, P128 bottom right: courtesy Don McMillan, PP130-131: TH/USN, P133: AC, PP134-135 all: TH/USN, P136: courtesy Bill Hannan, P138: HMSI, P139: TH/USN, P141: TH/USN, P142: TH/USN, P143: courtesy John Bolt, P144: TH/USN, P145: NMNA, PP146-147 all: TH/USN, P148-149 both: courtesy Stanley Vejtasa, P150: PK, P151: John McQuarrie, P153: TH/USN, P154 all: PK, P155: PK, P158 margin and top: PK, P158 top right: TH/USN, P160: NMNA, P162: PK, P163 both: PK, P164: NMNA, P166: Neil Mercer, P167: TH/USN, P170: TH/USN, P171: TH/USN, P172: NMNA, P174: TH/USN, P176: PK, P177: PK, P178: TH/USN, P179: TH/USN, P182: TH/USN, P183: USAF, P185: TH/USN, PP186-187 all: TH/USN, P188: courtesy Michael O'Leary, P190 margin: courtesy Malcolm Bates, P192 top: PK, P192 bottom: TH/USN, P193: courtesy Paul Gillcrist, P194: TH/USN, P195: TH/USN, P196: TH/USN, P197: TH/USN, PP198-199 all: courtesy Malcom Bates, P201: TH/USN, P202 top: HMSI, P202 bottom: TH/USN, P203 both: TH/USN, PP206-207: TH/USN, P209 top: courtesy Don McMillan, P209 bottom: NMNA, P210 both: PK, P214 top: AC, P214 bottom: courtesy Bill Hannan, P217: IWM, P219: Neil Mercer, P220: courtesy Nigel Ward, P223: Neil Mercer, P224: IWM, P226: TH/USN, P227: AC, P229: NMNA, P230: TH/USN, P231 margin: courtesy Michael O'Leary, P232: TH/USN, P234 margin: AC, P235: AC, P239 top: PK, P239 bottom: courtesy Morris Montgomery, P241: TH/USN, PP242-243: TH/USN, PP244-245: TH/USN, P246 top: TH/USN, P246 bottom: PK, P247: HMSI, P250: TH/USN, P251: TH/USN, P253: PK.

Acknowledgments
The author is particularly grateful to the following people for their kind help in the development of this book: Bill Barr, Malcolm Bates, George Blair, John Bolt, Jennifer Brattle, Eric Brown, Piers Burnett, James Cain, Shannon Callahan, Larry Cauble, Parnell Chapell, Bill Ciazza, M. S. Cochran, Ed Copeland, Jack Crascall, Robert Croman, Frank Cronin, Tracy Curtis-Taylor, Dale Dean, Claire Donegan, John Dunbar, Dewey Durnford, Robert Elder, Lynn Forshee, Mark Fox, Oz Freire, Frank Furbish, Richard K. Gallagher, George Gay, Tess Gay, Paul Gillcrist, Jack Glass, Hill Goodspeed, Stephen Grey, Mark Hanna, Bill Hannan, Thomas Harris, Susan Henson, Tony Holmes, Chris Hurst, Jan Jacobs, Benny L. James, Dave Jones, Hargi Kaplan, Margaret L. Kaplan, Neal B. Kaplan, Kelly Kinsella, Jack Kleiss, Craig Kodera, Thomas LaGrange, Casey Law, Alan J. Leahy, Jack Leaming, Elvin Lindsay, Bruce Linsday, Paul Ludwig, Mike Machat, Richard H. May, John McCrarrie, Judith A. McCutcheon, Richard 'Mac' McCutcheon, Rick McCutcheon, Dale McGhee, Hamilton McWhorter, Don McMillian, Neil Mercer, Steve Millikan, Nick Mongillo, Morris Montgomery, H.B. Moranville, Kevin Morris, Jeff Mulkey, Patrick J. Nichols, Steve Nichols, David B. Oates, Michael O'Leary, Merle Olmsted, Arnold W. Olson, Brian O'Rourke, Bruce Porter, Henry Sakaida, Doug Siegfried, Wayne Skaggs, David Smith, David Soper, Mark Stanhope, Danny Stembridge, John Strane, Nagel Sullivan, David Tarry, Hannah Tausz, Mark Thistlethwaite, Ed Toner, Stanley Vejtasa, Danny Vincent, Michelle Vorce, Alex Vraciu, Nick Walker, George Wallace, Nigel 'Sharkey' Ward, Sky Webb, Wilbur Webb, Pat Weiland, John Wellham, Wilbur West, Scott Whelpley, Andy Withers, Dennis Wrynn, Aaron Zizzo.
Grateful acknowledgment is made to the following for the use of their previously published material:
Airlife Publishing Ltd.: Excerpts from *Wings of the Navy*, by Capt. Eric Brown. Reprinted by permission of the author.
Arms and Armour Press: Excerpts from *Kamikaze*, by Raymond Lamont-Brown. Reprinted by permission.
Bourne, David: For the use of his poem *Parachute Descent*.
Causley, Charles: For the use of his poem *HMS Glory*.
Dell, Division of Random House: Excerpt from *Air Warriors*, by Douglas C. Waller.
Fitz Roy, Olivia: For the use of her poem *Fleet Fighter*.
Gay, Tess (Mrs George Gay): Excerpts from *Sole Survivor*, by Ens. George Gay USNR, published by Midway Publishers. Reprinted by permission of Tess Gay.
Hutchinson Publishing Group: Excerpt from *War In A Stringbag*, by Charles Lamb.
Larteguy, Jean and George Blond: Excerpts from Larteguy's edited version of Blond's description of a kamikaze attack, in Larteguy's *The Sun Goes Down*.
McGraw-Hill: Excerpt from *Daybreak For Our Carrier*, by Lt. Max Miller USNR.
Naval Institute Press: Excerpt from *The Divine Wind*, by Capt. Rikihei Inoguchi and Cdr. Tadashi Nakajima with Roger Pineau.
Turner Publishing Company (Paducah, KY): Excerpts from *U.S.S. Enterprise (CV-6)*.
Ward, Nigel: Excerpts from *Sea Harrier Over The Falklands*, published by Orion. Reprinted by permission of the author.
Wedge, John: For the use of his poem *Still No Letter*.

Bibliography
Bennett, Christopher, *Supercarrier*, Motorbooks International, 1996.
Bowman, Martin W., *Shades of Blue*, Airlife Publishing Ltd., 1999.
Brown, David, *Kamikaze*, Brompton Books Corp., 1990.
Brown, Capt. Eric, *Wings of the Navy*, Airlife Publishing Ltd., 1987.
Burgess, Lt. Cdr. Richard R., *The Naval Aviation Guide*, Naval Institute Press, 1996.
Caidin, Martin, *Golden Wings*, Bramhall House, 1960.
Chant, Christopher, *Warships of the 20th Century*, Tiger Books International, 1996.
Chesneau, Roger, *Aircraft Carriers of the World, 1914 to the Present*, Arms and Armour Press, 1992.
Chinnery, Philip, *Desert Airforce*, Airlife Publishing Ltd., 1989.
Clancy, Tom, *Carrier*, Berkley Books, 1999.
Congdon, Don, *Combat WWII Pacific Theater of Operations*, Arbor

House, 1958.

Costello, John, *Love, Sex & War*, Collins, 1985.

Eliot-Morison, Samuel, *The Two-Ocean War*, Atlantic-Little Brown,1963.

Farmer, James H., *Celluloid Wings*, Tab Books, 1984.

Garrison, Peter, and Hall, George, *Carrier Aviation*, Presidio Press, 1980.

Goldstein, Donald M., Dillon, Katherine V., and Wenger, J. Michael, *Pearl Harbor, The Way It Was*, Brassey's, 1995.

Harrison, W. A., *Swordfish Special*, Ian Allen, 1977.

Holder, Bill and Wallace, Mike, *F/A-18 Hornet A Photo Chronicle*, Schiffer, 1997.

Holmes, Tony, *Combat Carriers*, Motorbooks International, 1998.

Holmes, Tony, *Seventh Fleet Supercarriers*, Osprey Publishing Ltd., 1987.

Holmes, Tony, and Montbazet, Jean-Pierre, *World Super Carriers*, Osprey Publishing, Ltd., 1988.

Horsley, Lt. Cdr. Terence, *Find, Fix and Strike*, Eyre & Spottiswoode, 1945.

Humble, Richard, *Aircraft Carriers*, Michael Joseph Ltd., 1982.

Ireland, Bernard, *The Rise and Fall of the Aircraft Carrier*, Marshall Cavendish, 1979.

Isby, David C., *Jane's How To Fly and Fight in the F/A-18 Hornet*, HarperCollins Publishers, 1997.

Jackson, Robert, *Air War Korea 1950–1953*, Airlife Publishing Ltd., 1998.

Johnsen, Frederick A., *Douglas A-1 Skyraider*, Schiffer, 1994.

Johnston, Stanley, *Queen of the Flattops*, E. P. Dutton & Co., 1942.

Johnstone-Bryden, Richard, *HMS Ark Royal IV*, Sutton Publishing Ltd., 1999.

Keeney, Douglas, and Butler, William, *No Easy Days*, BKF, 1995.

Kelly, Orr, *Hornet*, Airlife Publishing Ltd., 1990.

Lake, Jon, *Grumman F-14 Tomcat*, Aerospace Publishing, 1998.

Lamont-Brown, Raymond, *Kamikaze*, Arms and Armour Press, 1997.

Lawson, Robert, and Tillman, Barrett, *Carrier Air War*, Motorbooks International, 1996.

Loomis, Robert D., *Great American Fighter Pilots of World War II*, Random House, 1961.

Lowry, Thomas P., and Wellham, John W.G., *The Attack on Taranto*, Stackpole Books, 1995.

McManners, Hugh, *Top Guns*, Network Books, 1996.

Melhorn, Charles M., *Two-Block Fox*, Naval Institute Press, 1974.

Mercer, Neil, *The Sharp End*, Airlife Publishing Ltd., 1995.

Miller, Lt. Max, *Daybreak for our Carrier*, McGraw-Hill, 1944.

Montbazet, Jean-Pierre, *Super Carriers*, Osprey Publishing Ltd., 1989.

Musciano, Walter A., *Warbirds of the Sea*, Schiffer, 1994.

O'Leary, Michael, *United States Naval Fighters of World War II in Action*, Blandford Press, 1980.

Poolman, Kenneth, *Escort Carrier 1941–1945*, Ian Allen, 1972.

Porter, Col. R. Bruce, with Hammel, Eric, *Ace!*, Pacifica Press, 1985.

Preston, Antony, *Aircraft Carriers*, Bison Books Ltd., 1982.

Reynolds, Clark G., *The Fast Carriers*, McGraw-Hill, 1968.

Reynolds, Clark G., *The Fighting Lady*, Pictorial Histories Publishing, 1986.

Ross, Al, *The Escort Carrier Gambier Bay*, Conway Maritime Press, 1993.

Shores, Christopher, *Air Aces*, Bison Books Ltd., 1983.

Silverstone, Paul, *US Navy 1945 to the Present*, Arms and Armour Press, 1991.

Smith, John T., *The Linebacker Raids*, Arms and Armour Press, 1998.

Smithers, A. J., *Taranto 1940*, Leo Cooper, 1995.

Spick, Mike, *McDonnell Douglas F/A-18 Hornet*, Salamander Books Ltd., 1991.

Steichen, Edward, *U.S. Navy War Photographs Pearl Harbor to Tokyo Bay*, Bonanza Books, 1956.

Sweetman, Bill, *Joint Strike Fighter*, MBI Publishing Company, 1999.

Taylor, Michael, *Naval Air Power*, Hamlyn Publishing, 1986.

Watts, Anthony J., *The Royal Navy*, Arms and Armour Press, 1999.

Weiland, Charles Patrick, *Above & Beyond*, Pacifica Press, 1997.

Wellham, John, *With Naval Wings*, Spellmount, 1995.

Wolff, Brian R., and Alexander, John, *From The Sea*, Reed International Books, 1997.

Wragg, David, *Carrier Combat*, Sutton Publishing, 1997.

Wragg, David, *Wings Over The Sea*, David and Charles, 1979.

Wrynn, V. Dennis, *Forge of Freedom*, Motorbooks International, 1996.

Index